P9-CQC-715

Learning Astronomy
by Doing Astronomy

Learning Astronomy
by Doing Astronomy

Collaborative Lecture Activities

Stacy Palen

WEBER STATE UNIVERSITY

Ana Larson

UNIVERSITY OF WASHINGTON

W. W. NORTON & COMPANY

NEW YORK • LONDON

W. W. Norton & Company has been independent since its founding in 1923, when William Warder Norton and Mary D. Herter Norton first published lectures delivered at the People's Institute, the adult education division of New York City's Cooper Union. The firm soon expanded its program beyond the Institute, publishing books by celebrated academics from America and abroad. By midcentury, the two major pillars of Norton's publishing program—trade books and college texts—were firmly established. In the 1950s, the Norton family transferred control of the company to its employees, and today—with a staff of four hundred and a comparable number of trade, college, and professional titles published each year—W. W. Norton & Company stands as the largest and oldest publishing house owned wholly by its employees.

Copyright © 2015 by W. W. Norton & Company, Inc.

All rights reserved.
Printed in the United States of America

Editor: Erik Fahlgren
Project Editor: Diane Cipollone
Copy Editor: Christopher Curioli
Editorial Assistant: Arielle Holstein
Managing Editor, College: Marian Johnson
Director of Production, College: Jane Searle
Media Editor: Rob Bellinger
Assistant Media Editor: Paula Iborra
Marketing Manager: Stacy Loyal
Design Director: Hope Miller Goodell
Photo Editor: Stephanie Romeo
Composition and Illustrations: Precision Graphics
Manufacturing: Sheridan Printing
Permission to use copyrighted material is included in the credits section of this book, which begins on page 153.

ISBN: 978-0-393-26415-9

W. W. Norton & Company, Inc., 500 Fifth Avenue, New York, NY 10110
www.wwnorton.com

W. W. Norton & Company Ltd., Castle House, 75/76 Wells Street, London W1T 3QT

1 2 3 4 5 6 7 8 9 0

Contents

About the Authors

Stacy Palen received her bachelor's degree from Rutgers University in 1993, and her PhD from the University of Iowa in 1998. Upon graduation, she spent four years as a post-doc/lecturer at the University of Washington, where she taught introductory astronomy 20 times in four years. This experience enabled her to really focus on students' conceptual difficulties with the course material, and experiment with ways to address them. Her astronomical research has focused on multi-wavelength studies of dying sun-like stars. She has also conducted research in the teaching and learning of astronomy, and served on multiple national committees that promote astronomy through education and public outreach.

Currently, she is a Full Professor at Weber State University in Ogden, Utah. In addition to the usual professorial duties, she also runs the Ott Planetarium, an astronomy outreach center that produces planetarium content for all ages. She lives on a small farm with her husband, two dogs, five dairy goats, five horses, and an ever-changing number of chickens.

Ana Larson first had a short career in the business world followed by being a stay-at-home mom. She then received a double bachelor's degree in physics and astronomy from the University of Washington in 1990 and a PhD in physics from the University of Victoria in 1996. Upon graduation, she joined the University of Washington, where she is now a senior lecturer in astronomy. She has taught introductory astronomy approximately twice a year for the past 17+ years, most frequently for the large lecture-based course, and has also taught advanced courses in astrophysics. Her development and thorough testing of introductory, lecture-related activities has been an on-going effort for over a decade. Having recognized that students need to see connections between what is read in the textbook, taught in class, and practiced in sections, she has created activities that are both relevant and challenging. Ana has also taught introductory astronomy online at Seattle Central College for the past 15 years and has incorporated many of her activities into the course.

Ana is the director of the Jacobsen Observatory on the University of Washington campus and guides undergraduates of all majors in their educational outreach efforts as they introduce the night sky to the public during viewing nights.

Preface

Dear Student,

The study of astronomy is filled with wonder, excitement, and surprises. Though sitting and listening to someone lecture about wonder, excitement, and surprises can be very educational, it is much more fun to discover the surprises yourself, create your own excitement, and find your own answers to all the questions you wonder about. We have designed the activities in this book to do exactly that: to place in front of you the data, concepts, and skills, and then guide you as you develop your own understanding.

Throughout this book we emphasize a hands-on, minds-on approach in which you will be challenged to re-create discoveries, verify scientific principles, and question your pre-existing notions. Sometimes this will be very challenging! It can be very hard to let go of a comfortable idea that you have held for a long time. It can be very hard to have the discipline to keep working, even when you don't always see how the project will end. But the investment of your time and effort will be worth it. You will be rewarded with a deeper understanding not only of the universe itself, but also of the methods astronomers use to explore the universe. You will begin to develop a feel for how astronomers ask questions, seek answers, and share results.

Each activity is designed and classroom-tested to illuminate a particular astronomical concept or principle. In some cases, such as discovering the expansion of the universe or a new planet around another star, you will be working with real data from real telescopes. In other cases, you will be using metaphors to explore inaccessible regions of space, like the area around a black hole. In every case, you will be required to engage with the material to develop the deepest understanding.

Experience is the best teacher. We have assembled these activities to give you real-world experience with the universe and the objects in it. As you work your way through the book, pause occasionally to marvel at how much we know about the universe and, perhaps more importantly, how much remains to be discovered and understood. We hope you will be inspired to ask new questions, collect your own data, and find your own answers.

Dear Instructor,

Studies of teaching and learning confirm that a hands-on approach is "hands-down" the best way to learn. When students actively engage with the material in a course, they make great strides in overcoming misconceptions, acquiring new knowledge, and building new skills. We have created this workbook by following the best practices of astronomical education research.

Each activity is built around a basic concept in astronomy and leads the student from a novice understanding to a deeper understanding through a guided interaction. Students may interact with astronomical data or with common metaphors for astronomical objects, learning to predict the behavior and properties of astronomical objects. All of these activities have been "field-tested" in classrooms of various sizes, with students of varying backgrounds. All have been found to be effective for students whether they are learning in a large university setting, an open-enrollment regional university, or in an online course.

We have made a special effort to balance the activities among the four areas commonly found in introductory astronomy courses: physics and observations, the Solar System, stars, and galaxies and cosmology. You will find as many activities to use in the last half of your course as there are for the first half. Because of this broad coverage, the workbook may be used either to accompany *Understanding Our Universe* or *21ˢᵗ Century Astronomy* or it may be used as a stand-alone learning package for your students, accompanying your usual lecture.

We have included enough activities that you have some choice, but not so many as to overwhelm the students. With 30 activities from which to choose, instructors will have sufficient activities for a typical introductory course to use at least one per week.

These activities are designed to be used in the classroom, require no special equipment or preparation (although some would be enhanced by showing images on a screen), and can be completed within an hour by most students. Students working in groups of two or three will make the most effective use of the activities, although it would also be possible for them to work alone. The sections of each activity are arranged in steps, guiding the students from initial knowledge-level questions or practice to a final evaluation or synthesis of what they have just accomplished. Students thus get practice thinking at higher cognitive levels.

Each activity has pre-activity and post-activity questions. The pre-activity questions have been designed to address some of the common misconceptions that students have, to relate familiar analogous terrestrial examples to the activity, and to act as a brief refresher in such things as scale factors, measurements, and basic mathematics review. These questions should act as mindsets for students as they transition from their other courses and activities to their learning of astronomy. The post-activity questions review the most important concepts introduced in the activity. Thus, the post-activity questions can be used as assessment tools, as reading quizzes, or to stimulate post-activity discussions. You might assign the pre- and post-activities through Norton Smartwork for students to complete before they come to class or you might use them with "clickers" in the classroom. You might incorporate them into your other assessments as a way to test retention of the concepts learned through the activity.

We authors have been using activities like these in our classroom for more than 15 years. "Activity day" is always a favorite for us and students alike. We repeatedly find great joy in watching students figure out difficult concepts for themselves, develop confidence in their data analysis, graphing, and mathematical skills, and push each other to greater feats of intellectual bravery as they grow to understand that the universe is not magic. It is not impenetrable. It does not take a special talent or type of mind to understand it. It takes careful, thoughtful study and a willingness to ask every question that comes to mind and then chase down the answers. We are confident that you will see the same changes in your students that we see in ours. Please do check in with us and let us know how these activities work for you!

Acknowledgments

We would like to thank our reviewers, whose feedback greatly helped improve the accuracy of this workbook: Charles Kerton, Iowa State University; Nathan Miller, University of Wisconsin-Eau Claire; Dwight Russell, Baylor University; Don Terndrup, The Ohio State University.

Name _____ Date _____ Section_____

Mathematical and Scientific Methods

Learning Goals

This activity reviews the mathematics that you may encounter in this course. This exercise will help you with tools such as working with logarithms, the small-angle formula, scientific notation, or scaling exercises, like those used to find the scale of a map, laboratory techniques concerning measurements, measurement uncertainties, and statistical analysis. If you struggle with this activity, plan a review with your classmates or instructor soon.

In the first six sections, you will review specific mathematical topics and laboratory techniques. These sections include explanations and practice problems. In the last section, you will pull multiple concepts together to analyze images of galaxies.

You will also:

1. Demonstrate knowledge of the essentials of mathematics through practice and review of:
   - scientific notation and powers of 10
   - algebra
   - logarithms
   - the small-angle formula
   - the use of scale factors and scaling
   - statistics and uncertainties in measurements
2. Describe the process of science and the scientific approach as personally experienced.

Step 1—Review of Mathematics

1. Algebra: Recall that algebraic equations can be solved for any term in the equation, as long as you remember always to perform each operation on both sides of the equals sign. If you divide one side by a term to isolate the unknown, you must divide the other side by the same term.

   a. Solve for a in terms of F and m: $F = ma$. _____

   b. Solve for m in terms of E and c: $E = mc^2$. _____

   c. Solve for x in terms of y: $y = 4x^2$. _____

2. Scientific Notation: In astronomy, numbers tend to be very, very large. It is inconvenient and confusing to write them out in long-hand, with lots of zeros. Scientific notation makes these numbers compact by keeping only the first few digits (often no more than three) in the number, and then keeping track of the place in a separate term.

   For example, suppose that your calculator gives the result of a calculation as 132,465.54. Most of these digits are not important; in this book, keep only the first three. This number has the 1 in the hundred thousands place, five decimal places to the left of the decimal point. Therefore, 132,465.54 can be more compactly written as 1.32×10^5, where the positive 5 keeps track of the hundred thousands place. A negative 5 would mean the hundred-thousandths place: $0.0000132 = 1.32 \times 10^{-5}$.

   a. Write in scientific notation:

      i. 3,042 = _____

      ii. 231.4 = _____

      iii. 0.00012 = _____

      iv. 0.0000000000667 = _____

   b. Convert from scientific notation:

      i. 4.2×10^{14} = _____

      ii. 4×10^{-11} = _____

3. Powers of 10: When multiplying or dividing in scientific notation, treat the numbers out front normally (multiply or divide the two numbers). Then, for multiplication, add the exponents on each "10." For example: $(3 \times 10^2) \times (6 \times 10^3) = 18 \times 10^5 = 1.8 \times 10^6$. For division, subtract the exponent in the denominator from the one in the numerator. For example: $(3 \times 10^2) / (6 \times 10^3) = 0.5 \times 10^{-1} = 5 \times 10^{-2}$.

   a. Multiply: $(3.1 \times 10^7) \times (3 \times 10^5)$ = _____

   b. Divide: $(1.496 \times 10^{11}) / (5.2 \times 10^{-3})$ = _____

   c. Simplify: $(3 \times 10^8)^2$ = _____

4. Logarithms: Logarithms are a way of solving for the exponent in an equation, so that if you had an equation that had 10^x in it, you could solve for the x. For our purposes, it is sufficient that you know how to carry out the operation on your calculator using the "log" button.

   Use the equation $M = m - 5 \times \log d + 5$ to find M for each pair of values of m and d in the table below.

m	d	M
2.0	10.0	
1.5	15.0	
−1.5	3.0	

Step 2—Small-Angle Formula

For angles smaller than 0.5 radian (about 30°), the sine of the angle, the tangent of the angle, and the angle are all approximately equal to one another. That is, $\sin \theta = \tan \theta = \theta$, when the angle is expressed in radians.

The tangent of an angle is the opposite side of the triangle divided by the adjacent side, so that for the triangle shown in **Figure 1.1**, $\tan \theta = s/d$. If the angle is small, then the tangent of the angle is equal to the angle, and this equation reduces to $\theta = s/d$. This equation is known as the **small-angle formula**, where θ is the angular size in radians, s is the actual (linear) size of the object being measured, and d is the distance to the object. The units of s and d must match. For example, if s is in meters, d must also be in meters. If s is in light-years, then d must also be in light-years. If we know the distance to an object and can measure its angular size, we can calculate its actual size. If we know the actual size and measure the object's angular diameter, we can determine its distance. This is a very important formula in astronomy.

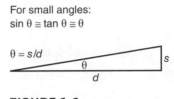

For small angles:
$\sin \theta \cong \tan \theta \cong \theta$

$\theta = s/d$

FIGURE 1.1

5. Use the small-angle formula to fill in the blank places in the following table.

ANGULAR DIAMETER (RADIANS)	ACTUAL DIAMETER (KILOMETERS)	DISTANCE (KILOMETERS)
0.01	3,480	
0.009		1.5×10^8
	3,474	3.84×10^5

Step 3—Scaling and Scale Factors

Every image, whether it is for astronomy, a snapshot for a scrapbook, or a sketch to use for a map, has a scale factor. On maps, the scale factor is always shown explicitly, usually in one corner. For example, 1 inch, as measured on the map, represents a different distance on the ground, perhaps 10 miles. In this case, the scale factor is 1 inch = 10 miles. In astronomy, this scale factor usually relates a measurement on the image to an angle on the sky; 1 millimeter (mm) = 20 arcseconds (arcsec), for example.

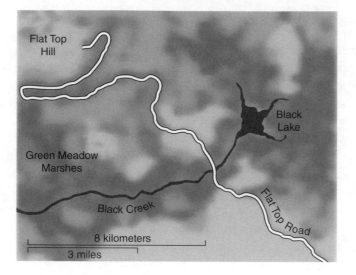

FIGURE 1.2

Use the map in **Figure 1.2** and the ruler given in **Figure 1.6** to answer the questions.

6. What is the measurement, in inches, that equals 3 miles? Convert that measurement so that you have 1 inch = _____ miles. (Fill in the blank.)

7. What is the measurement, in centimeters (cm), that equals 8 kilometers (km)? Convert that measurement, indicating that 1 cm = _____ km. (Fill in the blank.)

8. If a feature on the map is 3-cm wide, how many kilometers does that represent? _____ km

9. Use your scale factor found in question 7 to find the approximate length of Black Creek that is shown on the map. _____ km

10. Approximately how many kilometers is it to walk along Flat Top Road to get from the lower right-hand corner of the map to the end at Flat Top Hill? _____ km

11. If you walked at 2 kilometers per hour (km/h), how long would the trip take you? _____ hours

Step 4—Measurement Errors and Uncertainties

Every experiment has some uncertainty associated with it. Sometimes this is a random measurement uncertainty: you sometimes look at the ruler with your left eye, and sometimes with your right, causing you to read distances slightly longer or shorter each time. Sometimes this is an absolute measurement uncertainty: your ruler only reads down to millimeters and so cannot read to micrometers (μm; 1,000 times smaller). Sometimes this is a bias: it is easier to find really large asteroids, so at first, those were the only asteroids astronomers found. Knowing the uncertainty of a measurement is an important part of science. It helps us compare results from different experiments to one another and also helps us figure out how to improve our experimental methods by working on the sources of uncertainty to improve the precision of our measurements.

12. Using the ruler of **Figure 1.6**, measure the length of each object in **Figure 1.3**.

FIGURE 1.3

    a. Length: _____ cm

    b. Length: _____ cm

    c. Length: _____ cm

13. Which measurement will be the most precise; that is, have the least amount of uncertainty? _____ Which measurement will be least precise? _____

14. For which line could we justify the use of a micrometer capable of measuring accurately to two decimal places? _____

15. What are your estimates of your uncertainties (the possible error ranges in your measurements) for each line or shape? You can express this as a percentage or an actual number: 3 ± 0.5 cm, or 3 cm ± 17 percent, for example.

    a. _____

    b. _____

    c. _____

Step 5—Simple Statistics

The mean of a data set is the average, while the standard deviation gives a measure as to how spread out the data are. If we have enough data, we often make a graph to see if there is a relationship between variables. The two graphs in **Figure 1.4** plot the acceleration of a car with no engine problems (Figure 1.4a) and that of a car having fuel-injection issues (Figure 1.4b). The slope of the line is calculated by dividing the change in y by the corresponding change in x. Read these changes off of the axis labels by selecting two values of y and the corresponding values of x. The formula is: slope $= (y_2 - y_1) / (x_2 - x_1)$.

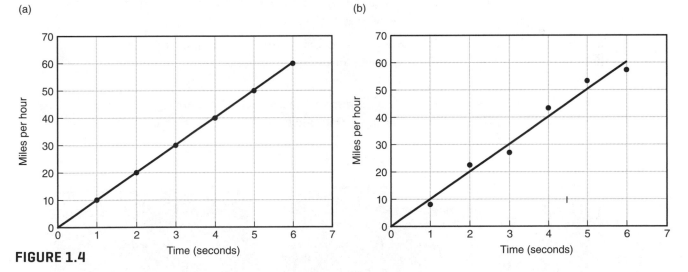

FIGURE 1.4

16. For the car with no engine problems, what is the slope of the line? _____

17. What are the units of the slope? _____

18. How fast would this car be going after 3.5 seconds? _____ mph

19. For the car with fuel-injection issues, what is the slope of the line? _____

20. How fast would this car be traveling after 3.5 seconds? _____

21. Give your estimate of the uncertainty in the slope of the line. _____

Step 6—The Scientific Method

In astronomy, the scientific process often begins with observation, and then follows a series of (not necessarily completely linear) steps, involving making educated predictions, testing them, and evaluating and retesting them. For example, the process might go this way:

> Observe :: Hypothesize (make an "educated guess," predict) :: Test :: Evaluate (reject, modify, or retain hypothesis) :: Form a new hypothesis if necessary :: Predict :: Retest :: Loop until hypothesis is retained, cannot be disproved at this time, and a new scientific theory or model is formed.

22. Describe briefly an example of the scientific process from your personal experiences.

Step 7—Combine Scaling, Small-Angle Formula, Scientific Notation, and Measurement Errors

If galaxies look to be the same type, then these galaxies may be assumed to be similar in actual size. Therefore, if one of the galaxies appears to be one-half the angular size of the other, then the apparently smaller galaxy is farther away—twice as far from us.

23. Each image in **Figure 1.5** is 600 arcsec on a side. Using the ruler in **Figure 1.6**, measure the length of the white line in one of the images and calculate the scale for all of the images:
1 cm = _____ arcsec.

Measure the diameters of the galaxies and express your answers in centimeters. Astronomers try to measure out as far as the spiral arms can be seen. Then, using your image scale, calculate the angular diameters in arcseconds.

(a)   (b)   (c)

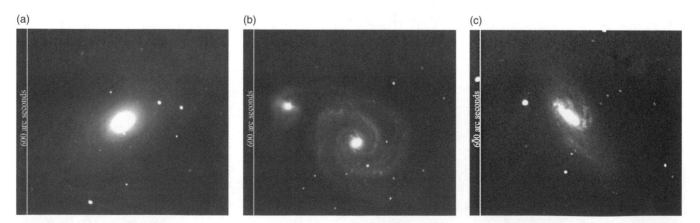

FIGURE 1.5

24. Diameters in centimeters: A _____ B _____ C _____

25. Angular diameters in arcseconds: A _____ B _____ C _____

To use the small-angle formula, we must express the angle in radians, rather than arcseconds. There are 206,265 arcsec in 1 radian (rad). Suppose a galaxy was 600 arcsec in angular size. Converting that to radians is easiest to visualize if we use ratios:

$$\frac{1 \text{ rad}}{206,265 \text{ arcsec}} = \frac{x \text{ rad}}{600 \text{ arcsec}}$$

To solve for x, we multiply both sides by 600 arcsec, and then flip the equation around:

$$600 \text{ arcsec} \times \frac{1 \text{ rad}}{206,265 \text{ arcsec}} = \frac{x \text{ rad}}{600 \text{ arcsec}} \times 600 \text{ arcsec}$$

$$600 \text{ arcsec} \times \frac{1 \text{ rad}}{206,265 \text{ arcsec}} = x \text{ rad}$$

$$x \text{ rad} = 600 \text{ arcsec} \times \frac{1 \text{ rad}}{206,265 \text{ arcsec}}$$

$$x = 2.91 \times 10^{-3} \text{ rad}$$

26. Follow the above example to convert the angular diameters you calculated above to radians.
A _____ B _____ C _____

27. The angular size of a galaxy is inversely proportional to its distance from us. How much farther away from us is the farthest galaxy compared to the nearest? _____

If we knew the actual size of these galaxies, then we could use the small-angle formula to find the distance. Because these galaxies are all spiral galaxies like the Milky Way, they are all approximately the same size as the Milky Way, so we can use the estimated diameter of the Milky Way—100,000 light-years—as our standard ruler. Manipulate the small-angle formula to solve for distance and use scientific notation for your answers.

28. Distance to galaxy A _____; B _____; and C _____.

29. Which galaxy has the most uncertainty in your measurement of its angular diameter? _____Which one has the least? _____ Explain your answer, using the appropriate techniques we have covered in this activity.

FIGURE 1.6

ACTIVITY 2

Astronomical Measurements: Examples from Astronomical Research

Learning Goals

In this activity you will explore the relationship between apparent brightness, luminosity, and distance and learn to manipulate more advanced equations used in astronomy. Specifically, you will learn to:

1. Apply the small-angle formula.

2. Distinguish between apparent magnitude and absolute magnitude and relate them correctly to the concepts of apparent brightness and luminosity.

3. Relate the ratio of distances to the brightness ratio for stars of equal luminosity.

4. Solve for the distance to a star using the parallax angle.

5. Find the absolute magnitude from the apparent magnitude and the distance.

6. Demonstrate proficiency in manipulating more advanced equations used in astronomical research.

Step 1—The Small-Angle Formula

Although the units may vary from one problem to another when you are using the small-angle formula to determine sizes, always use the same units for the distance to an object and the linear diameter of that object. Also, take care to use the correct form of the small-angle formula for the angular units. If θ is in radians, then $\theta = s/d$, where θ is the angular diameter, s is the linear diameter, and d is the distance. If θ is in arcseconds, then

$$\frac{\theta}{206,265} = \frac{s}{d}$$

1. Use the small-angle formula to fill in the missing information in **Table 2.1**. Be sure to fill in the units in the first column.

○ TABLE 2.1

Angular diameter, actual (linear) diameter, and distances for selected astronomical objects.

OBJECT	CAT'S EYE NEBULA	HORSEHEAD NEBULA	PLEIADES CLUSTER	GLOBULAR CLUSTER M71
Angular diameter (arcseconds)	20	300	6,000	430
Actual diameter (_____)				
Distance (light-years)	3,300	1,630	440	13,000

Step 2—Magnitudes

We measure the apparent brightness of an object in a logarithmic unit called a **magnitude**. Every five magnitudes corresponds to a change in brightness of 100, so that a magnitude 1 star is 100 times brighter than a magnitude 6 star. (Notice that the magnitude scale is "backwards": smaller numbers mean brighter stars.) **Table 2.2** shows the relationship between magnitude differences (diff) and apparent brightness ratios (ratio). From Table 2.2, we can find out how the brightness compares if we know the difference in apparent magnitudes.

2. Fill in the missing information in Table 2.2, using the pattern that every difference of five magnitudes corresponds to a factor of 100 in the brightness ratio.

Because every five magnitudes corresponds to a factor of 100 in brightness, each magnitude step corresponds to a factor of $(100)^{1/5}$, or 2.512 in brightness. In general, the brightness ratio is given by $= (2.512)^{m_2 - m_1}$, where $m_2 - m_1$ is the difference in magnitudes. For example, a magnitude 3 star and a magnitude 5 star have a difference in magnitudes of 2, so the magnitude 3 star is $(2.512)^2$ times brighter than the magnitude 5 star.

3. How many times brighter is a magnitude 5 star than a magnitude 6 star? _____

4. How many times brighter is a magnitude 7 star than a magnitude 10 star? _____

5. How many times brighter is a magnitude 0 star than a magnitude 5 star? _____

○ TABLE 2.2

The relationship between magnitudes and apparent brightness.

DIFF	RATIO	DIFF	RATIO	DIFF	RATIO	DIFF	RATIO	DIFF	RATIO
0	1	1	2.5	2	6.3	3	16	4	40
5		6	250	7	630	8	1,600		4,000
10	10,000	11	25,000	12		13	160,000	14	400,000
15	1,000,000								

Step 3—The Inverse Square Law

If two stars **have identical luminosities**, then the ratio of their apparent brightness is equal to the inverse of the ratio of the distances squared:

$$\frac{B_1}{B_2} = \frac{d_2^2}{d_1^2}$$

Notice that on the left, the brightness of object "1" is on top, but on the right, the distance to object "1" is on the bottom.

   If we know the ratio of the apparent brightness, we can figure out the relative distances. Suppose that star 1 and star 2 are two stars of the same type, so they have the same luminosity. But star 1 has an apparent brightness in our sky that is 230 times greater than that of star 2 (B_1/B_2 = 230). Star 1 must be nearer than star 2 because it appears brighter in our sky. How much farther away is star 2 than star 1? Substituting the brightness ratio of 230 into the equation above gives

$$230 = \frac{d_2^2}{d_1^2}$$

Multiply both sides by d_1^2, and then flip the equation around to get

$$d_2^2 = 230 \times d_1^2$$

To isolate the distance to star 2, d_2, we must take the square root of both sides:

$$\sqrt{d_2^2} = \sqrt{(230 \times d_1^2)}$$

$$d_2 = \sqrt{230} \times d_1$$

$$d_2 = 15d_1$$

So star 2 is 15 times farther away than star 1.

   6. Follow the above example to fill in the missing numbers and conclusions in **Table 2.3**.

Step 4—Parsec

The word *parsec* is an abbreviated form of **parallax arcsecond** and is a convenient unit for astronomy. Nearby stars appear to shift across the sky when observed from two different vantage points from the Earth as it orbits the Sun. The shift is measured as an angle, called the parallax angle, p. Once this parallax angle has been measured (in arcseconds), the distance in parsecs, d, is just the inverse of the angle:

$$d = 1/p$$

   7. Suppose a star has a parallax angle of 0.00126 arcsecond (arcsec). What is its distance in parsecs (pc)? _____

   8. Suppose a star has a distance of 4.3 pc. What is its parallax angle in arcseconds? _____
      Fill in column 4 of Table 2.4.

○ TABLE 2.3

Magnitudes, brightness ratios, and relative distances of pairs of stars.

SPECTRAL TYPE	STAR NAME AND (APPARENT MAGNITUDE)	MAGNITUDE DIFFERENCE	RATIO OF BRIGHTNESS	RATIO OF DISTANCES	CONCLUSION
O9.5 V	Mu Columbae (5.17) Zeta Ophiuchi (2.56)	2.61	$(2.512)^{2.61} = 11.1$	$\sqrt{11.1} = 3.33$	Mu Columbae is 3.3 times farther than Zeta Ophiuchi.
B3 V	Alkaid (1.85) Regulus (1.35)				
A1 V	Sirius (−1.46) Merak (2.37)	3.83	$(2.512)^{3.83} = 34$	$\sqrt{34} = 5.8$	Merak is 5.8 times farther than Sirius.
F0 V	Porrima (3.65) Alkalurops (4.31)	0.66	$(2.512)^{0.66} = 1.84$	$\sqrt{1.84} = 1.35$	Alkalurops is 1.35 times farther than Porrima.
G2 V	Rigel Kentaurus (−0.02) Sun (−26.72)				
K5 V	61 Cygni A (5.2) Kaffaljidhma C (3.47)				
M6 V	Wolf 359 (13.53) Ross 248 (12.29)				

The "V" following the spectral type of the star is the star's luminosity class. All of these stars are on the main sequence in an H-R diagram and are thus class "V."

Step 5—Absolute Magnitude

The magnitude that we have been using above is more precisely called the **apparent magnitude**, m_v. This is the magnitude the star appears to have from Earth. This magnitude does not tell us anything directly about the star itself—it could be faint in our sky because it is very far away or it could be faint in our sky because it is not luminous. The amount of light emitted at visible wavelengths is measure by its **absolute magnitude**, M_v. The absolute magnitude is related to both the distance, d, and the apparent magnitude, m_v:

$$M_v = m_v - 5 \log d + 5$$

The apparent magnitude can be observed directly from the ground. The distance can be found, for example, by parallax. Once the distance and the apparent magnitude are known, the absolute magnitude can be found, which is a measure of the total amount of energy emitted by the star.

9. Fill in the missing numbers in **Table 2.4** based on the given observed quantities for each star. The subscript V attached to the absolute and apparent magnitudes indicates measurements in the visible portion of the spectrum.

● TABLE 2.4

Observed and derived quantities for selected nearby stars.

STAR NAME	APPARENT MAGNITUDE, m_v	PARALLAX (ARCSECONDS)	DISTANCE (PARSECS)	ABSOLUTE MAGNITUDE, M_v
Sirius	−1.46	0.37921	2.64	1.43
Rigel	0.12	0.00378		
Betelgeuse	0.42	0.00655		
Deneb	1.25	0.00231		
Regulus	1.35	0.04113		
Bellatrix	1.64	0.01292		
Enif	2.40	0.00473		

10. The stars are listed according to their apparent magnitudes (their brightness in our sky). What is the order of the stars if you rank them according to distance, from nearest to farthest?

Nearest _____ _____ _____ _____ _____ _____ _____ Farthest

Is this order the same as when the stars were listed according to apparent magnitude?

_____

11. What is the order of the stars if you rank them according to absolute magnitude from lowest to highest?

Lowest _____ _____ _____ _____ _____ _____ _____ Highest

Is this order the same as when the stars were listed according to apparent magnitude?

_____

ACTIVITY 3

Where on Earth Are You?

Learning Goals

In this activity you will learn about the coordinate systems that are used on Earth and how our location on Earth and Earth's orbit around the Sun is related to the seasons. You should also be able to:

1. Recognize that the Sun and stars appear differently at different locations on Earth.

2. Summarize how these differences lead to seasons on Earth.

3. State where the seasons are most and least extreme on Earth and how this difference follows from the location of the Sun in the sky.

Step 1—Coordinates on Earth

Figure 3.1 shows Earth at two points during its orbit. One of these points is northern winter solstice, and the other is northern summer solstice.

1. On Figure 3.1, fill in the four blanks indicating whether it is summer or winter in the Northern or Southern hemisphere.

2. Five locations (Arctic Circle, Tropic of Cancer, Equator, Tropic of Capricorn, and Antarctic Circle) are labeled on Figure 3.1. Take a moment to study the locations and latitudes in each image.

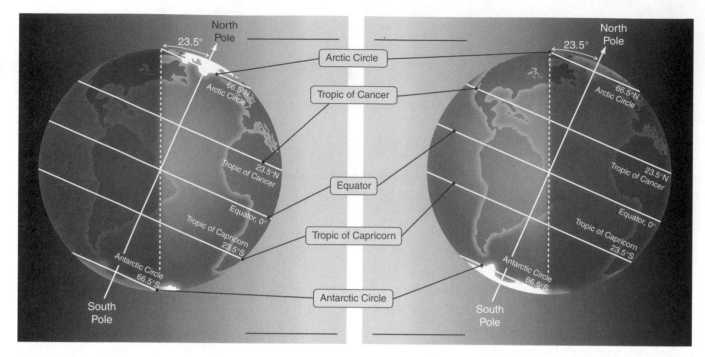

FIGURE 3.1

Step 2—The Celestial Sphere

Figure 3.2 shows a simplified view of the imaginary celestial sphere. Earth is at the center, and the shaded plane is the plane of the ecliptic: this is the path of the Sun along the celestial sphere through the year.

3. In Figure 3.2, notice the four special points along the ecliptic. Using the perspective of an observer in the Northern Hemisphere, label the position of the Sun on the summer solstice, the winter solstice, the vernal (spring) equinox, and the autumnal equinox.

4. Label the north and south celestial poles.

5. Label the celestial equator.

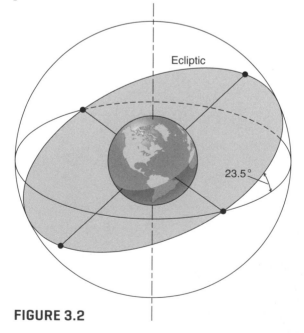

Ecliptic

23.5°

FIGURE 3.2

Step 3—Combining Perspectives

In **Table 3.1**, match the description of the sky with the location from which that observation would be seen. More than one location may apply to each description.

● TABLE 3.1

Descriptions of the sky at various locations on Earth

DESCRIPTION OF SKY	LOCATION ON EARTH
1. The Sun can be seen at the zenith twice during the year. _____	a. North Pole
2. North circumpolar stars are seen. _____	b. Tropic of Cancer
3. The Sun can be seen at the zenith only once during the year. _____	c. South of the Tropic of Cancer
4. The north celestial pole can be seen at the zenith. _____	d. North of the equator
5. All stars rise and set. _____	e. Equator
6. All northern stars are circumpolar. _____	f. South of the equator
7. Celestial poles are seen on the horizon. _____	g. North of the Tropic of Capricorn
8. South celestial pole is seen at the zenith. _____	h. Tropic of Capricorn
9. All southern stars are circumpolar. _____	i. South Pole
10. The ecliptic is directly overhead at local noon on the northern hemisphere's summer solstice. _____	
11. The ecliptic is directly overhead at local noon on the northern hemisphere's winter solstice. _____	
12. The Sun is at the zenith on the equinoxes. _____	
13. The Sun fails to rise above the horizon between the northern hemisphere's spring and autumn equinoxes. _____	
14. The Sun passes through the zenith at winter solstice. _____	

Step 4—Putting It Together

6. For an observer in the Northern Hemisphere, describe how the maximum altitude of the Sun in the sky changes throughout the year.

7. Considering the location of the Sun in the sky, where on Earth do you expect to find the most extreme seasons? Why?

ACTIVITY 4

Studying the Phases of the Moon from a Privileged View

Learning Goals

Understanding the phases of the Moon requires visualizing the Earth-Moon-Sun system in three dimensions. In this activity, you will develop this ability by learning how to:

1. Successfully replicate the motions of the Earth and Moon, as well as their positions with respect to the Sun at each lunar phase.

2. Explain the continuity of the Moon phases worldwide.

3. Use an Earth-Moon figure to disprove a common misconception that Moon phases are caused by Earth's shadow.

4. Correctly order the phases of the Moon.

Step 1—Understanding the Rotation and Revolution

Take a full sheet of paper and crush it into a ball. This will represent Earth. Find something in your backpack to represent the Moon. The object should be about one-quarter the size of the crushed-paper Earth. If necessary, you can crush one-quarter of a sheet of paper to represent the Moon. Study the polar graph on the last page of this activity (see **Figure 4.5**). You want to make Earth rotate through 1 day while, at the same time, moving the Moon in its orbit the correct number of degrees each day. It will be easiest to do this if the models of Earth and the Moon are sitting on a guide. The Moon moves nearly 12° in its orbit every 24 hours (360° divided by a rounded-off 30 days in a month). For simplicity, consider all locations to mean those for viewers in the Northern Hemisphere—those who see the Moon located in the southern part of the sky as it crosses the meridian.

Place the polar graph sheet on the desk in front of you. Imagine the sunlight coming from the front of the classroom. Orient your paper so that the "Day 0 & 30" location on the graph is toward the "Sun." Set Earth on the center of the graph paper. Hold the Moon at "Day 0 & 30"— we will start with a new Moon, the same geometry as shown in **Figure 4.1**.

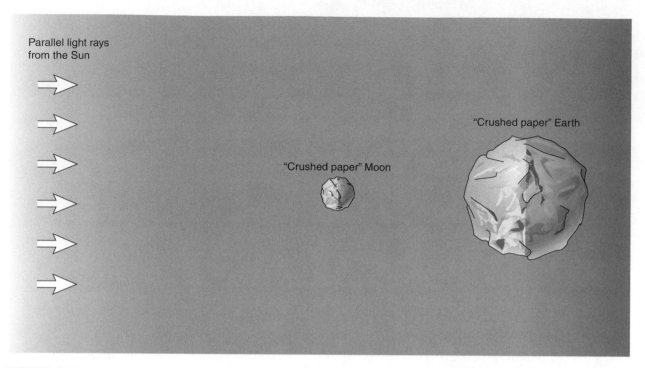

Parallel light rays
from the Sun

"Crushed paper" Earth

"Crushed paper" Moon

FIGURE 4.1

1. Starting with a new Moon, rotate the paper Earth once. If this is an accurate model, do all locations see approximately a new Moon (that is, if we could actually "see" a new Moon)? **Yes or No**

2. Each time Earth rotates, 1 day passes, and the Moon moves 12°. Move the Moon counter-clockwise in its orbit, and again rotate Earth. Keep stepping through this process (rotate Earth once, move the Moon 12°) until you get to the first quarter Moon (7.5 rotations of Earth). Do all locations on Earth see a first quarter Moon, or very close to this phase, over the course of one rotation of Earth? **Yes or No**

3. Move the Moon and rotate Earth until the alignment is Sun-Earth-Moon (Moon on the "Day 15" mark). What is the phase of the Moon when the system is in this configuration?
   _____

4. Do all locations on Earth see this phase, or very close to this phase, over the course of one rotation of Earth? **Yes or No**

5. Continue another 7.5 days. What is the phase of the Moon? _____

6. Briefly summarize your answers for questions 1–4, explaining what phases of the Moon are observed at different locations on Earth over just one complete rotation.

Step 2—Phases and Earth's Shadow

A common misconception is that the phases of the Moon are caused by Earth's shadow. **Figure 4.2** depicts the orbit of the Moon around Earth. The Sun is far off the paper to the left. If Earth's diameter in this image were 2.5 cm, Moon's orbit would have a radius of 75 cm, and the Sun would be more than three football fields away.

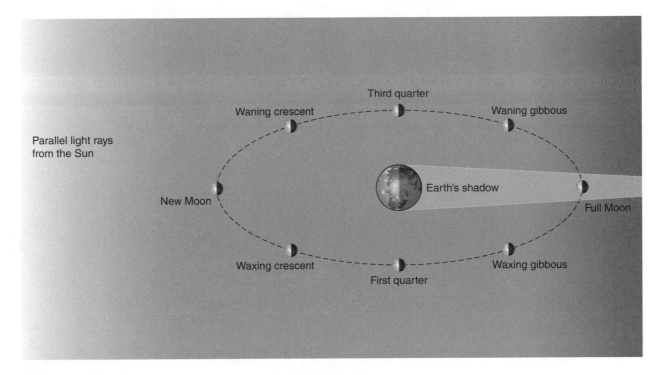

FIGURE 4.2

7. Earth's shadow is shown on Figure 4.2 as it would look when the Moon was full. If the phases of the Moon were caused by the Earth's shadow falling on it, would we ever see a full Moon? Explain your answer, bringing in why the inclination of the orbit of the Moon would matter.

8. Keeping the Earth's shadow in the direction and shape shown, could you show how any of the other phases could be caused by the shadow of Earth? How or why not?

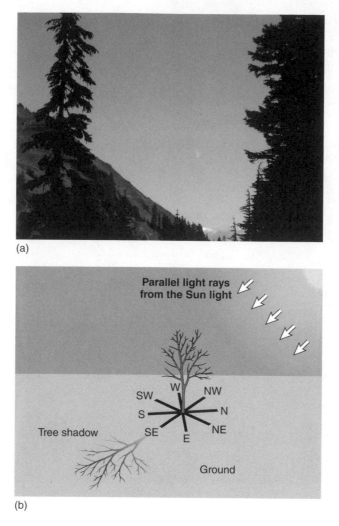

(a)

(b)

FIGURE 4.3

Just one actual observation is needed that negates this "shadowed Moon" idea in order to disprove the hypothesis. **Figure 4.3a** shows an observation that does just that. The phase is first quarter and the Sun is still up, located in the NW. The Moon is in the SW sky. Remember that the Moon is very far away, and the Sun is almost 400 times farther away than the Moon.

9. Shadows always point in the opposite direction from the Sun. Because the Sun is in the NW in Figure 4.3a, the shadows of the trees in Figure 4.3b must point toward the _____ direction.

10. The shadow of the Moon must point to the _____ direction.

11. The shadow of Earth must point toward its_____ direction

12. State how this observation disproves the "shadowed Moon" hypothesis.

Step 3—Understanding the Order of Moon Phases

Figure 4.4 shows the phases of the Moon in random order. Starting with "K" (a new Moon), put the phases in order, from new Moon to full Moon and back to new again.

k											k

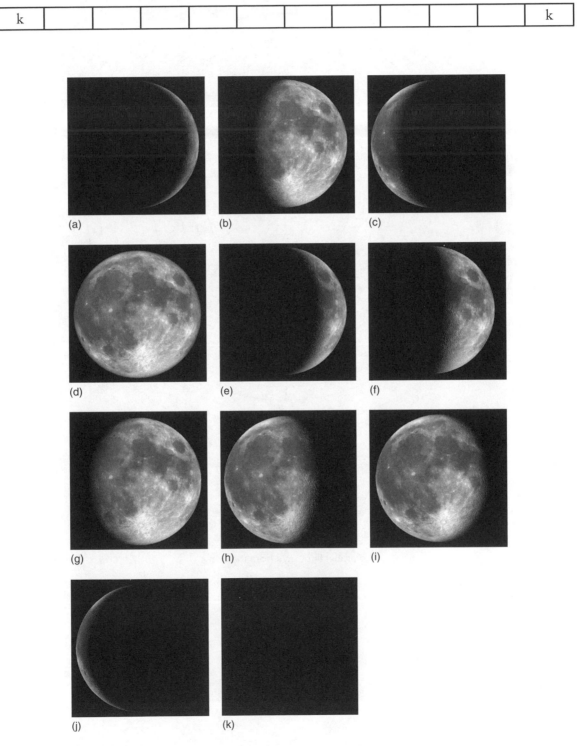

(a) (b) (c)

(d) (e) (f)

(g) (h) (i)

(j) (k)

FIGURE 4.4

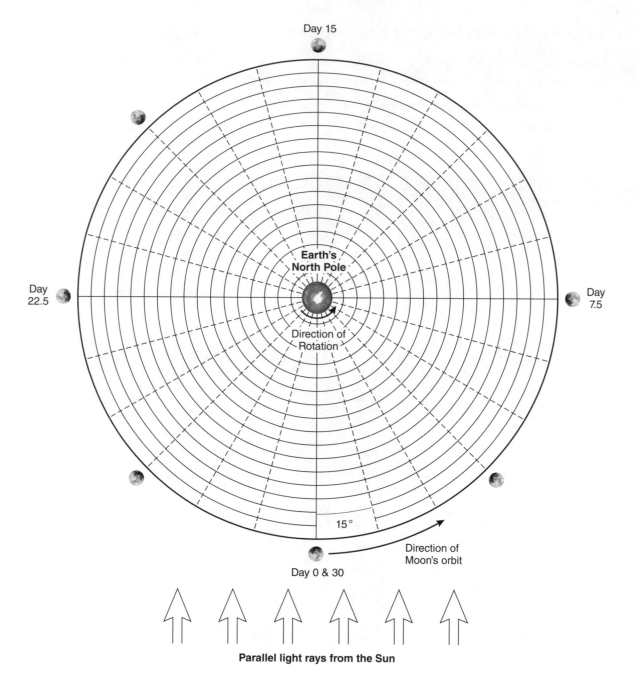

FIGURE 4.5

ACTIVITY 5

Altitudes of Objects on the Meridian at Your Location

Learning Goals

In this activity you will learn to connect the altitudes of celestial objects as they cross the celestial meridian to your latitude on Earth. In addition to helping you develop your abstract reasoning visualization skills, this activity will also help you:

1. Reproduce the altitudes of the north celestial pole, celestial equator, a star, and the Sun (at solstices and equinoxes) on the meridian for your location.

2. Explain why there are different seasons throughout the year.

3. Explain how the stars appear to move through the sky and how the motion of the stars differs when viewed from different latitudes on Earth.

Step 1—Objects on the Meridian

Figure 5.1 shows a person looking up at the celestial sphere. The large, outer circle represents the celestial sphere, with the northern horizon on the left. For any observer, the **celestial meridian** runs from north to south through the zenith. The altitude of the north celestial pole above the northern horizon is the same as a person's latitude. Work out the following questions assuming the observer is you, and thus located at your latitude.

1. Draw a line from the center of this celestial sphere (marked by a dot) to the following features of the celestial sphere, labeling each one, and giving each angle along your meridian from either the northern or the southern horizon, as appropriate. The altitude in degrees from the northern and southern horizons are shown.

   a. the angle to the celestial north pole

   b. the celestial equator

   c. the altitude of the Sun at the summer solstice

   d. the altitude of the Sun at the winter solstice

   e. the altitude of the Sun at the equinoxes

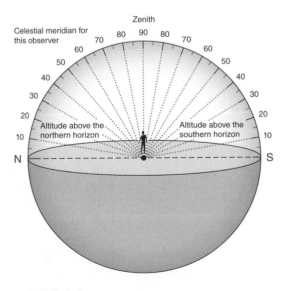

FIGURE 5.1

2. You are observing a star that is the same number of degrees above the celestial equator as your latitude. What is the altitude of that star when it crosses the meridian?

3. You spot a star that is right at the southern horizon as it crosses the meridian. How many degrees is that star below the celestial equator?

4. Examine the location of the Sun on the meridian at the times of the solstices and equinoxes. What does its altitude at these times have to do with the reason we experience seasons?

5. How does the motion of the star that is the same number of degrees above the celestial equator as your latitude differ from that of a star on the southern horizon?

6. Where will the north celestial pole be located if you traveled to the North Pole? What if you traveled to the equator?

7. All stars in the northern half of the celestial sphere are circumpolar from the North Pole. Explain what the motion of the stars at the equator would be. What feature on the celestial sphere would pass through your zenith at the equator?

8. Write a short summary of what you have learned here. In what ways will you be able transfer this knowledge to the actual night sky?

ACTIVITY 6

Working with Kepler's Laws

Learning Goals

In this activity, you will learn about Kepler's geometric model of planetary orbits and learn to:

1. Determine the properties of an ellipse.

2. Apply these properties to planetary orbits.

Step 1—Kepler's First Law

Kepler's first law states that the orbits of the planets are ellipses with the Sun at one of the foci.

1. On **Figure 6.1**, clearly label the following parts of an ellipse:

    a. focus

    b. semimajor axis

    c. minor axis

    d. center

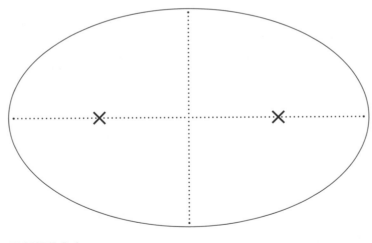

FIGURE 6.1

2. Eccentricity is a measure of the degree of "flattening" of the ellipse. Mathematically, the eccentricity of an ellipse is defined as the distance from a focus to the center of the ellipse divided by the length of the semimajor axis. Calculate the eccentricity of the ellipse in Figure 6.1. (Hint: Use the dotted lines and the number of "dots" as the units.)

3. A circle is a special ellipse, one with both foci at the same point. The eccentricity of a circle is 0. The value of the eccentricity of an orbit may run from 0 to almost 1. Estimate the eccentricities for the ellipses in **Figure 6.2**.

a. _____

b. _____

c. _____

4. In your own words, state how to determine the eccentricity of an ellipse. If you'd like, you may use a figure to express your thoughts.

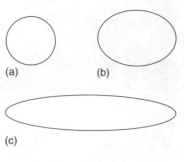

(a)　　　　(b)

(c)

FIGURE 6.2

Step 2—Kepler's Second Law

Kepler's second law, which states that as a planet moves around its orbit it sweeps out equal areas in equal times, can be difficult to visualize. It may be easier to consider this simpler version:

A planet travels faster when nearer to the Sun and slower when farther from the Sun.

Examine **Figure 6.3**, which shows a planet, D, orbiting the Sun, A, in a counterclockwise direction. Match the following terms with the letter that identifies the location in the figure:

5. focus _____

6. aphelion _____

7. perihelion _____

8. increasing speed: _____ to _____

9. decreasing speed: _____ to _____

10. planet has greatest speed _____

11. planet has lowest speed _____

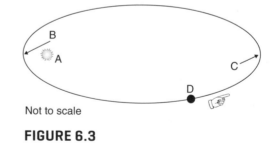

Not to scale

FIGURE 6.3

12. Now, state Kepler's second law in your own words, and give your reasoning for why planets behave this way.

Step 3—Kepler's Third Law

Kepler's third law relates the time it takes a planet to go around the Sun (the period, P) to the semimajor axis, a, of the orbit. If we assume that the mass of the Sun is much greater than any of the planets (which it is), and measure P in years and a in astronomical units (AU), the simplified relationship (formula) is

$$P^2 = a^3$$

13. According to Kepler's third law, all orbits with the same semimajor axis have the same period. The two orbits in **Figure 6.4** have the same value for a. How do their values of P compare?

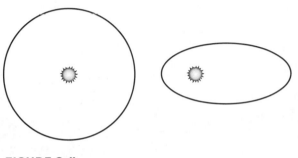

FIGURE 6.4

14. Explain how this is possible when the orbits look so different.

15. Fill in **Table 6.1** by calculating the missing data for the planets of our own Solar System. Be sure to include only the correct number of significant figures.

● TABLE 6.1

The semimajor axes and periods of the planets of the Solar System.

PLANET	SEMIMAJOR AXIS (AU)	PERIOD (YEARS)
Mercury	0.39	
Venus		0.62
Earth	1.0	
Mars		1.9
Jupiter	5.2	
Saturn		29
Uranus	19	
Neptune		165

Suppose a planet is discovered around a star like the Sun, with a period of 0.2 year.

16. What is the semimajor axis of its orbit? _____

17. How does this compare to the orbits of the planets in our Solar System?

ACTIVITY 7

Extraterrestrial Tourism

Learning Goals

In this activity, you will apply what you have learned about gravity to find out more about the exciting new discoveries of extrasolar planets (also called exoplanets). The Kepler mission, which involves a satellite in an Earth-trailing orbit (one that is slightly larger than Earth's), has already discovered close to 1,000 new exoplanets. By using the information for five of these recently discovered exoplanets, you will learn to:

1. Map the distances of the exoplanets from their stars onto the Solar System's scale.

2. Compare the hypothetical living conditions to those on Earth, based upon the distances the exoplanets are from their stars.

3. Using ratios, compare what one would weigh on one of the exoplanets to one's weight on Earth.

4. Discuss what it would be like under the gravitational effects of a planet other than ours.

5. Relate how Kepler's laws and Newton's law of gravity allow us to compare possible conditions on exoplanets to those on Earth.

Step 1—An Overview of Exoplanets

For ease of comparison, **Table 7.1** lists the mass of each exoplanet in "Earth masses"; that is, the mass of the exoplanet has been divided by Earth's mass, 5.97×10^{24} kilograms (kg). Similarly, the radius of each exoplanet is given in "Earth radii": It has been divided by Earth's radius, 6.37×10^6 meters.

○ **TABLE 7.1**

The original data.*

EXOPLANET DESIGNATION	ORBITAL PERIOD (DAYS)	ORBITAL PERIOD (YEARS)	ORBITAL DISTANCE (AU)	MASS COMPARED TO EARTH'S	RADIUS COMPARED TO EARTH'S
Kepler 3b	4.9	0.013	0.06	26	5
Kepler 4b	3.2	0.009	0.04	24	4
Kepler 8b	3.5	0.01	0.05	190	16
Kepler-34(AB)b	290	0.79	0.86	69	9
Kepler-35(AB)b	130	0.36	0.51	40	8

*Most values have been rounded off to no more than two significant digits.

1. Summarize how the masses and radii of the exoplanets compare to those of Earth.

2. Which exoplanet would be most *unlike* Earth, and in what way?

Step 2—Applying Kepler's Laws

The fourth column of Table 7.1 shows the orbital distances of the extrasolar planets in astronomical units (AU). Each of these planets orbits a star *similar in mass to the Sun*; this fact allows us to use Kepler's laws for these exoplanets.

3. Compare the distances of these exoplanets from their parent stars with the distance Earth is from the Sun (1 AU).

4. If these five exoplanets were actually *in our Solar System*, what would the planet order be?

5. Mercury's distance from the Sun is about 0.4 AU; that of Venus is about 0.7 AU; and that of Mars is about 1.5 AU. Assuming that the five exoplanets of Table 7.1 were part of our Solar System, mark and label their locations on **Figure 7.1**.

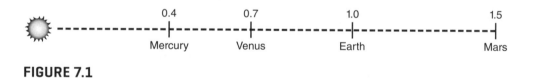

FIGURE 7.1

6. Pick one of the exoplanets as "your exoplanet" and write its name here:

7. What do you think the temperature would be like on your extrasolar planet?

8. Would you feel comfortable living on an extrasolar planet with that temperature?

Step 3—Applying Newton's Laws of Universal Gravitation

Because astronomy deals with very large numbers, it is usually more convenient as well as more insightful to work with ratios. For example, we find that Jupiter's mass is around 318 times that of Earth's, and its radius is about 11 times that of Earth's. On initial consideration, we might guess we would weigh an awful lot more on Jupiter if, in fact, it had a hard surface we could stand on. But, by using ratios we find that we would weigh a little less than three times as much.

9. Your weight is Earth's gravitational force on you (and, equally, your pull on it). Compare what you would weigh on each of the exoplanets versus your weight here on Earth and fill in **Table 7.2**, using the example for Kepler 3b as your guide. Set up Newton's law of gravity for the planet and for Earth. Combine your equations in a ratio, and then solve for the ratio of the forces. Only then should you plug in the numbers from the table. Show your work.

⦿ TABLE 7.2

Exercise for question 9.

EXOPLANET DESIGNATION	MASS COMPARED TO EARTH'S	RADIUS COMPARED TO EARTH'S	RATIO OF WEIGHTS/FORCES (EXOPLANET VERSUS EARTH)
Kepler 3b	26	5	$26/5^2 \approx 25/25 \approx 1$
Kepler 4b	24	4	
Kepler 8b	190	16	
Kepler-34(AB) b	69	9	
Kepler-35(AB) b	40	8	

10. Picking the same exoplanet as you did for question 6, would you be able to walk around the surface or jump up and down? What other experiences could you have at your "new" weight?

11. Briefly describe how an understanding of Kepler's laws and Newton's law of gravity allows us to compare conditions on exoplanets to those on Earth.

ACTIVITY 8

Light and Spectra

Learning Goals

In this activity, you will explore the properties of continuous and line emission. These properties are fundamental to our understanding of temperatures, compositions, luminosities, and velocities of astronomical objects. By working through this activity, you will be able to:

1. Explain how the temperature of an incandescent lightbulb affects the intensity and colors of the observed spectrum.

2. Compare the observed continuous spectrum to a series of Planck (blackbody) curves.

3. Examine emission spectra of five elements, noting the patterns and intensities of the lines.

4. Identify three "unknown" elements.

5. State the significance of each element having its own unique spectral signature of emission lines.

Step 1—Incandescent Bulbs and the Continuous Spectrum

The *first image* (shown by your instructor or displayed by other means) contains actual spectra from an incandescent bulb, starting at its brightest at the top and ending with its dimmest at the bottom. These are "continuous" spectra because the colors run smoothly across the wavelengths. Use colored pencils—or work with shading with regular pencils—to reproduce the brightest spectrum in **Figure 8.1a** and the dimmest in **Figure 8.1b**. If you use a pencil, label each wavelength with the color that appears in the spectrum at that wavelength.

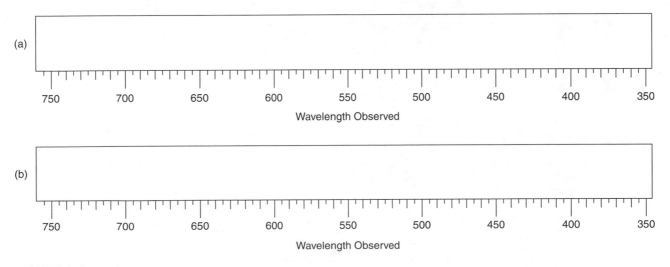

FIGURE 8.1

1. Based on your experience, is a dim bulb hotter or cooler than a bright bulb? _____

2. How does the temperature affect the bulb's intensity?

3. How does the temperature of the bulb affect the colors?

4. There are obviously colors missing in the spectrum when the bulb is at its dimmest compared to the spectrum of the bulb when it is at its brightest. Which colors are missing?

5. What do your answers to questions 2, 3, and 4 imply about the intensity of the light and the colors observed when viewing a hot versus a cool incandescent bulb?

Step 2—Relating the Results to Planck (Blackbody) Curves

Figure 8.2 shows a series of Planck curves that represent the intensity of light produced for a range of wavelengths and for four different temperatures. These curves were generated mathematically and conform to the usual practice of having short wavelengths on the left and long wavelengths on the right for the x-axis. This is opposite to what is displayed in the continuous and emission spectra being examined and reproduced here.

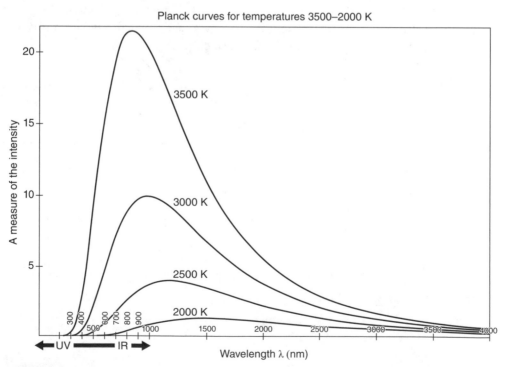

FIGURE 8.2

6. Do these theoretical curves support the summary you provided in question 5? Explain your answer, bringing in the intensity of the wavelengths within the visible part of the spectrum (300–700 nm).

7. The peaks in intensity for these curves occur in which region of the spectrum? Can we see light at these wavelengths?

The melting point for tungsten, the element for the filament of a bulb, is about 3700 K, which means that the filament probably does not get much hotter than about 3500 K. Examine **Figure 8.3** that gives an expanded view of the curve for 2000 K.

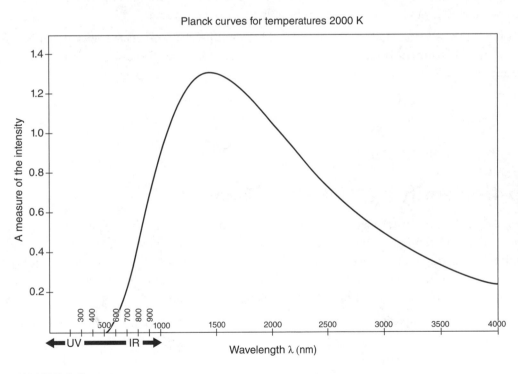

FIGURE 8.3

8. Assume that the temperature of the filament in the bulb at its dimmest was around 2000 K and explain how we were still able to see some red in the spectrum. If we could see at infrared wavelengths, would the bulb appear brighter or dimmer?

9. There are many stars that have surface temperatures in the 3500–2000 K range. Would they appear bright or dim to us? What kind of telescope would astronomers use in order to view these stars at their brightest wavelengths?

10. Where would be the best location for that telescope?

Step 3 — Emission Spectra

The *second series* of images (shown by your instructor or displayed by other means) to review contains the true-color emission spectra of five elements. In **Figure 8.4**, use colored or regular pencils to reproduce these spectra, placing the lines at the correct wavelengths. Indicate the brightness of the line by making wider lines on your paper for brighter lines in the spectrum. Be careful to note at what wavelength each color appears and the intensity of each line.

11. Compare the spectra that you sketched. Did any of the elements have the same emission spectrum? Comment on both the similarities you observed in these spectra and also the differences.

Step 4 — Identifying "Unknown" Elements

The *final series* of images (shown by your instructor or displayed by other means) to examine involve "unknown" elements shown in actual color, intensity, and spacing of the lines. Consider the emission spectra you sketched in Figure 8.4 to be the laboratory reference spectra and the unknowns to be from a newly discovered star. The pattern of the lines, colors, and spacing are important here since we may not know the wavelengths in advance.

12. Based on your comparisons between the known and unknown elements, what elements are present in this star?

13. From your results, summarize how spectra can be used to find the composition of a gas. Include in your summary the significance of each element having a unique spectrum.

Element names: a. _____ b. _____

c. _____ d. _____ e. _____

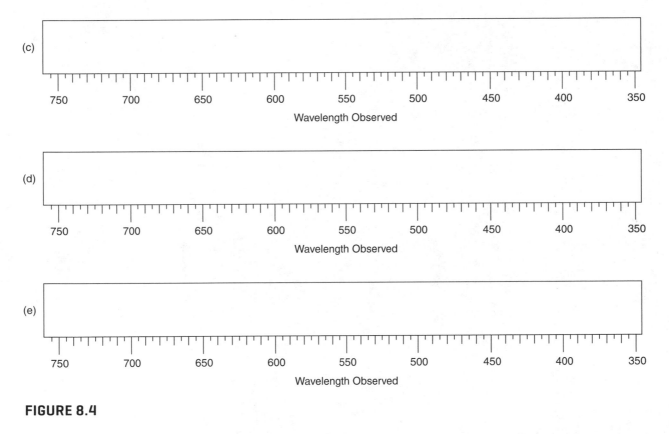

(a)

750   700   650   600   550   500   450   400   350

Wavelength Observed

(b)

750   700   650   600   550   500   450   400   350

Wavelength Observed

(c)

750   700   650   600   550   500   450   400   350

Wavelength Observed

(d)

750   700   650   600   550   500   450   400   350

Wavelength Observed

(e)

750   700   650   600   550   500   450   400   350

Wavelength Observed

FIGURE 8.4

ACTIVITY 9

Where to Put the Telescope?

Learning Goals

In this activity, you will learn about the impact of the atmosphere on astronomical observations and observatories. After completing this activity you should be able to:

1. Summarize how Earth's atmosphere affects decisions about which telescopes are built and where they are located.

2. Distinguish among telescopes that can be used on the ground versus those that must be located in space in order to observe celestial objects at specific wavelengths.

3. Explain the advantages and disadvantages of ground-based telescopes and telescopes in orbit.

Step 1—Earth's Atmosphere

Figure 9.1 shows the complete electromagnetic spectrum. Study this figure closely. Note for which wavelength regions the atmosphere is opaque, partially transparent, or completely transparent. Consider for which regions satellite observatories are required and those regions where ground-based observatories are used. Think also about why a number of large telescopes are located on top of high mountains in Hawai'i and Chile.

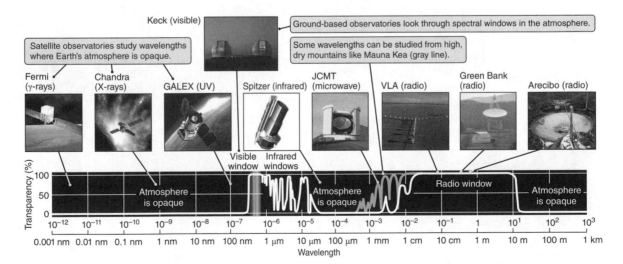

FIGURE 9.1

1. What does it mean for Earth's atmosphere to be opaque in a given band of electromagnetic radiation?

2. For which types of electromagnetic radiation is Earth's atmosphere completely opaque?

3. For which types of electromagnetic radiation is Earth's atmosphere partially opaque?

4. For which types of electromagnetic radiation is Earth's atmosphere completely transparent?

Step 2—Telescopes

5. There are telescopes for every band of the electromagnetic spectrum. Which types of telescopes will be able to detect flux from celestial objects if the telescopes are located on Earth?

   a. gamma-ray telescopes

   b. X-ray telescopes

   c. ultraviolet (UV) telescopes

   d. visible-light telescopes

   e. infrared telescopes

   f. radio telescopes

6. Which types of telescopes *must* be placed and operated in space, well above Earth's atmosphere?

   a. gamma-ray telescopes

   b. X-ray telescopes

   c. UV telescopes

   d. visible-light telescopes

   e. infrared telescopes

   f. radio telescopes

7. Briefly describe at least two advantages and two disadvantages for ground-based versus space-based telescopes.

ACTIVITY 10

51 Pegasi: The Discovery of a New Planet

Learning Goals

In this activity you will determine a planet's orbital period and the radial velocity amplitude of the parent star and then use these data with equations derived from Kepler's and Newton's laws in order to find the radius of the orbit and a lower limit on the mass of the planet. By comparing this planetary system to our Solar System you will also gain an appreciation for why astronomers were so surprised and skeptical when this discovery was first announced in 1995. When you have completed this activity you should be able to:

1. Apply Kepler's and Newton's laws to find orbital and physical characteristics of an exoplanet.

2. Compare an extrasolar planet to the more familiar planets of our Solar System.

Step 1—Observations

Figure 10.1 shows the observed radial velocities of the star 51 Pegasi over a period of about 33 days. The data were obtained by measuring the Doppler shift for the star using the Doppler formula. We find out what the shift in the wavelength is compared to the rest wavelength and multiply that number by the speed of light.

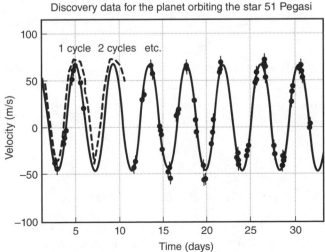

FIGURE 10.1

Step 2—Finding the Properties of the Planet

1. A period is defined as one complete cycle; that is, where the radial velocities return to the same position on the curve but at a later time. How many cycles did the star go through during the 33 or so days of observations?

Number of cycles = _____

2. What is the period, P, in days of one complete cycle? (Number of days for these observations divided by number of cycles.)

Period = _____ days

3. What is P in years? (Hint: Divide the period in days by the number of days in a year; the answer will be a decimal number smaller than 1.)

P = _____ years

4. What is the uncertainty in your determination of the period? That is, by how many days or fractions of a day could your value be wrong? (This is a number that you decide.)

Uncertainty = _____ days

5. What is the amplitude, K? To find this, take one-half of the value of the full range of the velocities.

K = _____ m/s

6. How accurate is your determination of this value?

Uncertainty = _____ m/s

7. We use a simplified form of Newton's version of Kepler's third law for determining the mass of the planet. The equation we use is

$$\frac{M_{planet}}{M_{Jupiter}} = \left(\frac{P}{12}\right)^{1/3} \times \left(\frac{K}{13}\right)$$

Period P should be expressed as a fraction of a year, and amplitude K should be expressed in meters per second (m/s). Twelve years is the approximate orbital period for Jupiter, and 13 m/s is the magnitude of the "wobble" of the Sun due to Jupiter's gravitational pull. The answer we get is the ratio of the mass of the planet (M_{planet}) to the mass of Jupiter ($M_{Jupiter}$). Put in your values for P and K and calculate the mass of this new planet in terms of the mass of Jupiter.

M_{planet} = _____ $M_{Jupiter}$

8. We assume that the parent star is 1 solar mass and that the planet is much, much less massive than the star (the case with our Solar System). Calculate the distance this planet is away from its star, in astronomical units (AU), using Kepler's third law, $P^2 = a^3$. Again, P is expressed as a fraction of a year, and a represents the astronomical units. Solve for a.

a = _____ AU

Step 3—Making Sense of the Calculations

9. Compare your results with the set of published results shown in **Table 10.1**. Make your comparisons quantitative by calculating the percentage differences:

$$\frac{\text{Your value} - \text{Published value}}{\text{Published value}} \times 100\%$$

O TABLE 10.1

Comparison of your results with the published results for the planet.

CHARACTERISTIC	PUBLISHED VALUE	MY VALUE	PERCENT DIFFERENCE
Mass	$0.45\ M_{\text{Jupiter}}$		
Period (P)	4.233 days		
Amplitude (K)	56.83 m/s		
Distance from star (a)	0.0527 ± 0.0030 AU		

10. Where would the new planet fit in if it were in our Solar System? **Figure 10.2** depicts the distances of the planets from the Sun to scale: Mercury, 0.4 AU; Venus, 0.7 AU; Earth, 1.0 AU; Mars, 1.5 AU; Jupiter, 5.2 AU.

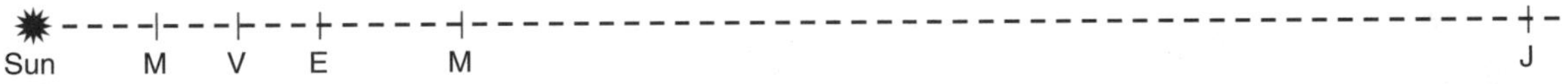

FIGURE 10.2

11. Science is based on the ability to predict outcomes. However, nothing prepared astronomers for the characteristics of this "new" solar system. Consider the mass of this planet as well as its distance from its star. Why was the discovery such a surprise when compared to our Solar System?

12. If this actually is a planet, is it possibly hospitable to life? Comment on what the environment would be like on this planet.

ACTIVITY 11

Ranking Task for Formation of Planetary Systems

Learning Goals

You will rank in order the events that take place during star and planet formation. You should visualize the birth of a star and planets in steps that begin from a giant, cool, rotating molecular cloud and end with fully formed planets orbiting a genuine star. You will also learn to:

1. Correctly order the stages of the formation of a planetary system.

2. Identify the stages where overlapping of events occurs.

3. Recount the modern theory of planetary system formation.

Step 1—Rank the Steps

The cards in **Figure 11.1** list the descriptions of the various stages of star and planet formation. Place them in the correct order. If any stages are ongoing or occur simultaneously, indicate that by adding an asterisk next to the letter.

Earliest _____ _____ _____ _____ _____ _____ _____ _____ Latest

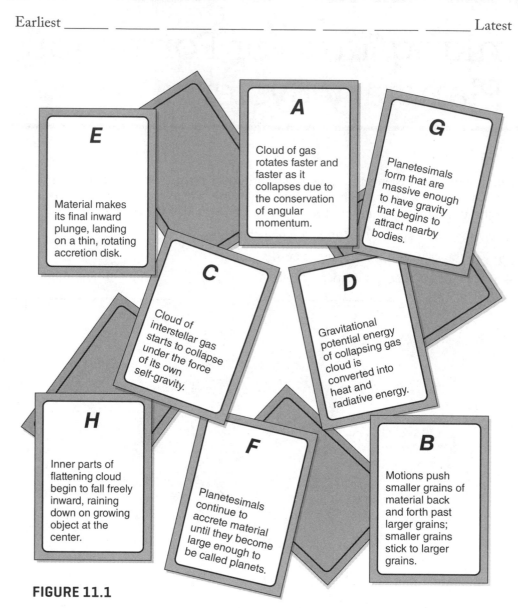

E

Material makes its final inward plunge, landing on a thin, rotating accretion disk.

A

Cloud of gas rotates faster and faster as it collapses due to the conservation of angular momentum.

G

Planetesimals form that are massive enough to have gravity that begins to attract nearby bodies.

C

Cloud of interstellar gas starts to collapse under the force of its own self-gravity.

D

Gravitational potential energy of collapsing gas cloud is converted into heat and radiative energy.

H

Inner parts of flattening cloud begin to fall freely inward, raining down on growing object at the center.

F

Planetesimals continue to accrete material until they become large enough to be called planets.

B

Motions push smaller grains of material back and forth past larger grains; smaller grains stick to larger grains.

FIGURE 11.1

Step 2—Recount the Theory of Planetary System Formation

Based on the ranking of the stages, write a summarizing paragraph that "tells the story" of how a planetary system forms.

ACTIVITY 12

Age Dating the Lunar Surface Using Cratering Statistics

Learning Goals

In this activity you will find the relative ages of regions on the Moon by comparing the distribution of crater sizes in different regions. This is a method that planetary scientists still use today. Whereas the scientists have a computer program designed to do this, you will use two much smaller contrasting regions of the Moon to work with manually. During this activity you will:

1. Set criteria for measuring the sizes of craters on the Moon.

2. Determine the distribution of sizes for both heavily and lightly cratered regions.

3. State the relative ages of the heavily and lightly cratered regions of the Moon and the actual range of ages derived.

4. List advantages and disadvantages between finding relative versus actual ages of surfaces.

5. Evaluate the effectiveness of this method for dating cratering events in the Solar System.

Step 1—Overview of the Lunar Surface

Figure 12.1 is a mosaic of images of the Moon taken during first quarter on July 26, 2012.

1. Compare the lower-right part of the mosaic of images to the upper-left part. Make a general statement about the number and densities of the craters.

2. Extend this summary to what the numbers and densities of the craters imply about the relative ages of those two regions of the Moon.

3. Combine your answers to the previous two questions and formulate a working hypothesis.

FIGURE 12.1

Step 2—The Cratered Highlands

We will start testing your hypothesis with the high-density region. Using the centimeter scale and circle sizes included in **Figure 12.2** count the number of craters of each approximate size in **Figure 12.3**, and insert the results in the left-hand side of **Table 12.1**. Figure 12.3 is an image of a small part of the *cratered highlands* region on the Moon. Calculate the image scale by dividing the size given in the small inset image in kilometers by the size of the crater in the full image in centimeters. Your scale will then be the number of kilometers per centimeter. For reference the craters noted are Albategnius (136 km diameter), Arzachel (97 km diameter), and Alpetragius (40 km diameter).

Be sure to keep track of which craters you have already measured so you don't count them twice. Count even the faintest craters you can see and those that overlap other craters. Because you are restricting your measurements to what are known as "binned" sizes, round off your sizes and enter the counts in the table. If you do not locate a crater of a given diameter, enter 1 for number counted.

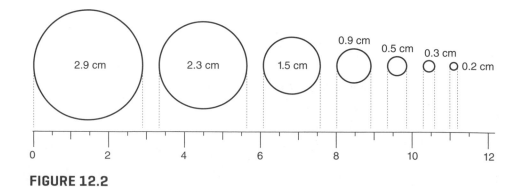

FIGURE 12.2

Albategnius (136 km)

Arzachel (97 km)

Alpetragius (40 km)

FIGURE 12.3

● TABLE 12.1

Crater counts for two regions of the Moon.

CRATER COUNTS FOR THE CRATERED HIGHLANDS REGION		CRATER COUNTS FOR THE UPPER MARE REGION	
ACTUAL SIZE (KM)	**NUMBER COUNTED**	**ACTUAL SIZE (KM)**	**NUMBER COUNTED**
180+		180+	
150		150	
120		120	
100		100	
80		80	
60		60	
40		40	
30		30	
20		20	
10		10	

Step 3—The Upper Mare Region

Using the same criteria you used above, fill out the right-hand side of Table 12.1 by measuring **Figure 12.4**, which shows a small part of the *upper mare* region on the Moon. Again, you will need to find the image scale using the small inset image of the figure for reference. The craters noted in this image are Theaetetus (25 km), Aristillus (55 km), and Archimedes (83 km).

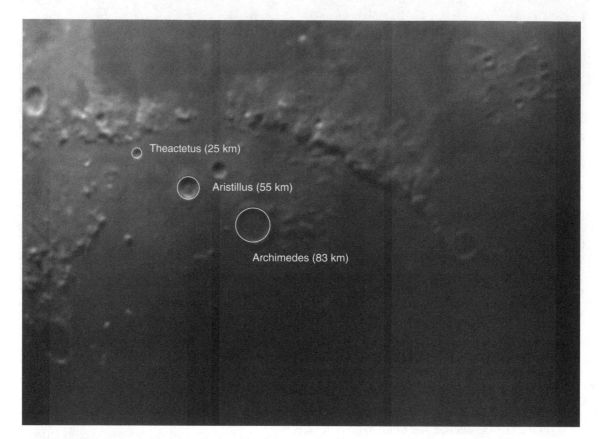

FIGURE 12.4

Step 4—Graphing the Data

The graph shown in **Figure 12.5** uses a logarithmic scale for the *y*-axis and a linear scale for the *x*-axis. The diagonal lines show the surface ages in billions of years based on crater counts, from 3.9 billion years (giga-years; Gyr) down to 3.7 billion years. These ages were found through radiometric dating of Moon rocks brought back by *Apollo* astronauts. The sunlit part of the images you measured cover just about 1 million square kilometers, so you can compare your results to published results, as the diagonal lines represent counts versus diameter over 1 million square kilometers as well. Graph the number of craters you counted versus actual size for both the cratered highlands region and the upper mare region. Use different symbols for the two regions.

Symbol used for cratered highlands: _____

Symbol used for upper mare region: _____

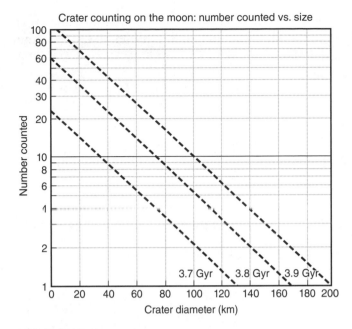

FIGURE 12.5

Step 5—Reflection Questions

4. What do the numbers and densities of the craters imply about the relative ages of these two regions of the Moon? Do your results support your working hypothesis? Comment, making sure you include the effects of uncertainties in identifying and including all craters.

5. Comparing crater densities on a terrestrial world easily leads to statements of relative ages, but relating crater densities to actual ages for the Moon required our sending astronauts there to bring back rocks to be dated based on radioactive isotopes. Based on your results, what are the approximate ages or age ranges for the cratered highlands region and the upper mare region?

6. Give an evaluation of this method for finding the relative ages and for estimating the actual ages of the surfaces of the Moon. What are the advantages and disadvantages in finding relative versus actual ages?

7. Conclude with this question and support your answer: Based on complete crater counts of the surfaces of the Moon, would we be justified in applying the actual ages of its surfaces to all worlds in the Solar System that have solid surfaces?

ACTIVITY 13

Comparison of Storms on Jupiter and Earth

Learning Goals

Through the use of *Voyager I* spacecraft observations of the Great Red Spot on Jupiter, you will not only learn about the wind conditions in Jupiter's upper atmosphere but also will:

1. Find the circumference of the Great Red Spot.

2. Calculate the rotation period and speed of the Great Red Spot.

3. Compare the speed of the Great Red Spot and its lifetime to one of the strongest storms on Earth.

4. Compare the results with published values, and consider sources of errors.

Step 1—*Voyager 1* Imagery

Before you begin, study the images taken by the *Voyager 1* spacecraft from January 6 to February 3, 1979, shown in **Figure 13.1**. These images were taken through a blue filter, one every Jupiter day, as the spacecraft was approaching the giant planet from 58 million to 31 million kilometers (km) away during that time frame. By taking the images at the same Jupiter local time, the Great Red Spot appears to remain stationary, while the belts and zones move across the image (both ways). Notice the action around the Great Red Spot as well as the belts and zones.

Follow the white feature (marked by an "X") around one complete rotation on the outer edge of the Great Red Spot, as shown in Figure 13.1, and judge if it is a good feature to use. You will need to calculate the period of rotation and also the speed of the material at this distance from the center of the Great Red Spot using this feature.

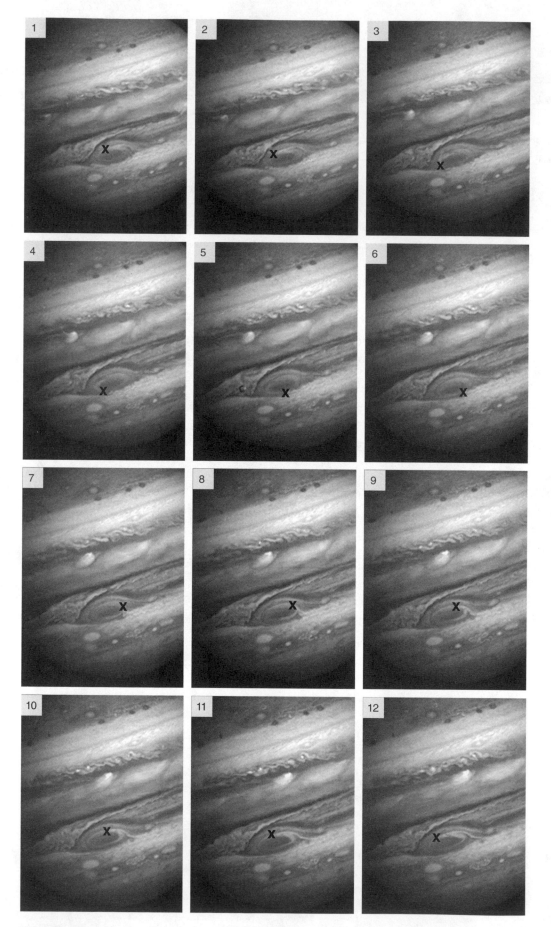

FIGURE 13.1

Step 2—Circumference of the Great Red Spot

Figure 13.2 shows an image of the Great Red Spot with an ellipse overlaid. For comparison, two "Earths" are inserted inside the Great Red Spot, emphasizing its great size. The circumference of Earth, measured by passing through the poles, is about 40,000 km.

FIGURE 13.2

1. Estimate the circumference of the ellipse in kilometers.

   _____ km

2. Figure 13.1 shows only a portion of the 60 images *Voyager I* took over 28 days from January 6 to February 3, 1979. Approximately how many hours, on average, passed between each image being taken? (Show your work here.)

   _____ hours

3. How many hours are covered by the 12 images in Figure 13.1, all taken in sequence? (Show your work here.)

   _____ hours

4. Divide the circumference of the ellipse you calculated above in kilometers (_____ km) by the number of total hours covered by the 12 images (_____ hours) to find the speed of the Great Red Spot at the distance of the white spot from its center. (Show your work here.)

   _____ km/hr

5. This is the approximate rotation **speed** in 1979. What is the rotation **period** at this distance from the center of the Great Red Spot? (Show your work here.)

   _____ days

Step 3—Comparison to One of the Strongest Cyclones on Earth

Since a couple of Earths fit into the Great Red Spot, it is meaningless to compare sizes for the Great Red Spot and a cyclone on Earth. However, we can compare speeds and duration.

The cyclone shown in **Figure 13.3** developed from a tropical depression on April 22, 1991, in the Bay of Bengal. Just before it reached landfall, it had wind speeds of 160 mph (258 km/h), the equivalent of a category 5 hurricane. After making landfall near Chittagong, Bangladesh, the storm weakened and dissipated by April 30, a total lifetime of just over a week.

FIGURE 13.3

6. Quantitatively compare the maximum speed of the cyclone to your value for the Great Red Spot. Comment on the comparison.

7. Current measurements of the velocity of the outer parts of the Great Red Spot are 610 km/h [D.S. Choi et al., *Icarus*, 188:35–46 (2007)] or the range 430–680 km/h (see http://missionjuno.swri.edu/jupiter/great-red-spot). Quantitatively compare your results to the range of values and comment.

8. What do you think are possible sources of error in your measurements?

ACTIVITY 14

The Clearing and Herding of Saturn's Ring Particles

Learning Goals

In this activity you will use Kepler's and Newton's laws to analyze the behavior of ring patterns in the rings of Saturn. After completing this activity you will be able to:

1. Apply Kepler's third law to sections of the ring system of Saturn.

2. Summarize the process by which a single moon will clear a gap in a ring system like Saturn's.

3. Summarize how the two shepherd moons manage to keep the particles that orbit between their orbits tightly confined within a narrow ring.

Step 1—Applying Kepler's Laws to Rings and Satellites

Figure 14.1 shows an image of part of Saturn's rings taken by the *Cassini* spacecraft. The rings are made up of millions of icy chunks. These ring particles follow Kepler's laws, each in its own orbit. From this image:

1. Which set of particles will be traveling fastest? (Check one.)

_____ A _____ B _____ C

2. Which set will be traveling the slowest? (Check one.)

_____ A _____ B _____ C

FIGURE 14.1

Figure 14.2 shows a *Cassini* image of Pandora, the F Ring, Prometheus, and the outer part of the A Ring of Saturn.

3. Rank the following objects in order of orbiting Saturn the fastest (1) to orbiting Saturn the slowest (4):

   _____ Pandora   _____ A Ring   _____ F Ring   _____ Prometheus

4. Explain your logic in ranking the objects.

Figure 14.3 depicts a possible approach of the space shuttle to a satellite in Earth orbit (Figure 14.3a). (Earth is at the bottom of the figure.) The shuttle has to capture the satellite to prevent it from falling out of orbit and burning up in Earth's atmosphere. The satellite is orbiting at a height of 250 km. The space shuttle is getting ready to adjust its orbit in order to catch up to the satellite, and there are three possible approaches (Figure 14.3b).

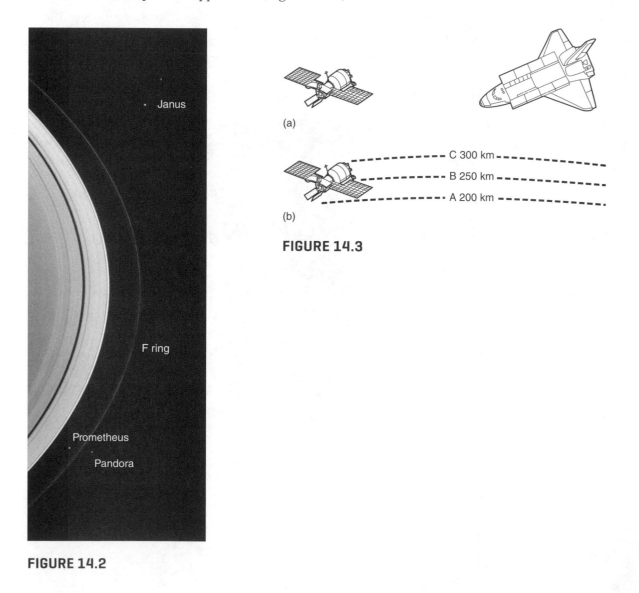

FIGURE 14.3

FIGURE 14.2

5. On which approach—A, B, or C—will the shuttle have the same speed as the satellite?
_____

6. On which approach—A, B, or C—will the shuttle be traveling slower than the satellite?
_____

7. That leaves a height of _____ where the shuttle will be traveling faster than the satellite.

8. A first-time commander of the space shuttle decides to orbit at 250 km and simply burn the shuttle's thrusters to catch up to the satellite rapidly. A mission specialist takes exception to this decision, stating that the space shuttle must catch up to the satellite from a lower orbit. Explain why the mission specialist is correct.

Step 2—How a Single Moon Clears a Gap in the Rings

Figure 14.4 shows a moon and two ring particles. The inner ring particle will have a slightly higher speed than the moon, as indicated by the longer arrow in the direction of its orbit, and the outer ring particle will have a slightly slower speed than the moon, as indicated by the shorter arrow in the direction of its orbit. The moon will have a strong gravitational effect on both ring particles.

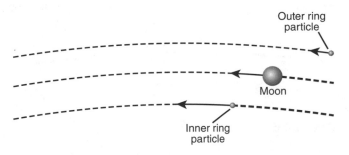

Action of a single moon on ring particles inside and outside its orbit.

FIGURE 14.4

9. *Draw arrows* that represent the direction in which the moon's gravity will act on these two particles. Label these arrows to distinguish them from the arrows representing the speeds of the particles.

10. Just after a particle loses speed, will it move "down" or "up"; that is, closer or farther from the planet? _____

11. Conversely, just after a particle gains speed, will it move "down" or "up"; that is, closer or farther from the planet? _____

12. Will the inner ring particle accelerate or decelerate due to the interaction? _____

13. Will the outer ring particle accelerate or decelerate due to the interaction?
_____

14. Now, *draw arrows* on Figure 14.4 that indicate the direction each particle will go in its orbit when affected by the single moon. Label these arrows to distinguish them from the arrows representing the gravitational forces.

15. As a particle falls "down" toward the planet, it will gain a little speed. This will stabilize it in a new, lower orbit. The outer particles, however, _____ speed as they move away from the planet, so they stabilize in a new, _____ orbit.

16. Summarize the process by which a single moon will clear a gap in a ring system like Saturn's.

Step 3—How Two Moons Shepherd Ring Particles

Figure 14.5 shows a ring interacting with two moons. The inner shepherd moon will have a slightly higher speed and the outer shepherd moon will have a slightly lower speed than the ring particles, as indicated by the sizes of the arrows pointing in the direction of the moons' orbits. We once again need to consider Kepler's third law.

17. Do the stray ring particles move faster or slower than the outer shepherd moon? _____

18. Do the stray ring particles move faster or slower than the inner shepherd moon? _____

19. The moons have a strong gravitational effect on the stray ring particles. *Draw arrows* on Figure 14.5 that represent the direction in which each moon's gravity will act on the stray ring

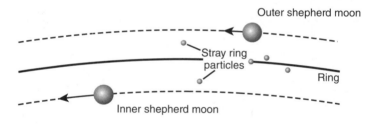

FIGURE 14.5

particle closest to it. Label these arrows to distinguish them from the arrows representing the speeds of the moons.

20. Now, consider your answers from the previous section, and *draw arrows* on Figure 14.5 that indicate the direction each stray ring particle will go in its orbit after being affected by the shepherd moon closest to it. Label these arrows to distinguish them from the arrows representing the gravitational forces.

21. Summarize how the two shepherd moons manage to keep the particles that orbit between their orbits tightly confined within a narrow ring.

ACTIVITY 15

Moons of the Giant Planets

Learning Goals

In this activity, you will compare major moons of the giant planets, both against one another and against Earth's Moon. This is an introduction to "comparative planetology," which is useful for finding commonalities among the bodies of the Solar System. As you work on this activity you will:

1. Compare and contrast the surface features of Jupiter's moon Ganymede with those of the Moon.

2. Discuss evidence of resurfacing and cratering on four moons of Saturn.

3. Compare and contrast surface features of Saturn's moon Titan to similar ones on Earth.

4. Compare the evidence for the reprocessing of part of Neptune's moon Triton.

Step 1—Jupiter's Moon Ganymede

Figure 15.1 shows the Moon and Ganymede side by side.

1. Name two surface features that Ganymede appears to have in common with the Moon.

Moon mean radius = 1737 km

2. In what ways are these features similar? In what ways must the origins of the features on Ganymede be different?

Ganymede mean radius = 2634 km

FIGURE 15.1

63

Step 2—Satellites of Saturn

Saturn has dozens of moons. **Figure 15.2** shows four of these moons: Enceladus, Dione, Rhea, and Mimas. These images are not to scale, but the mean radius is given for each moon.

Enceladus mean radius: ~250 km
Dione mean radius: ~560 km
Rhea mean radius: ~760 km
Mimas mean radius: ~200 km

FIGURE 15.2

Study these images and mentally note the amount of cratering, unusual terrain, coloring (shading), and so on of each moon. Are any of these surface features uniform among these moons? Envision the probable past evolution of each of these moons and answer the following questions.

3. Which moon(s) experienced resurfacing? Describe the evidence.

4. Are there areas of saturation cratering on any of these four moons? Which one(s) and on what part?

5. In general, are there regions on any of these moons where one could assume the surface is younger or older compared to regions of Earth's Moon? Explain your reasoning.

Step 3—Saturn's Majestic Moon: Titan

Even though Titan is a moon and is about 1.5 times the radius of the Moon, it is more interesting to compare Titan's surface features with those of Earth. Earth's average surface temperature is about 290 K; Titan's is about 94 K. The *Cassini* spacecraft, which orbits Saturn and passes by its moons, used infrared instruments and radar to map most of Titan's surface. There are a large number of features that have analogies on Earth. We investigate the most intriguing—lakes and tributary networks.

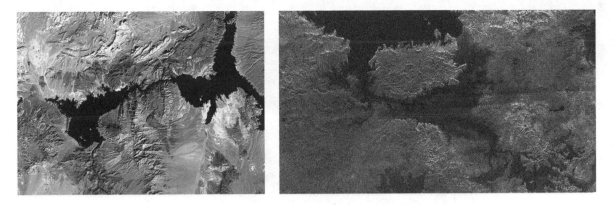

FIGURE 15.3

6. The images in **Figure 15.3** are from two different worlds. Figure 15.3a shows a pair of lakes on Earth, while Figure 15.3b shows a lake located in the polar region on Titan. For the pair of images shown in Figure 15.3, list a few things that look very similar and a few that are quite different.

Step 4—Triton, Largest Moon of Neptune

The *Voyager* spacecraft found evidence of the reprocessing of Triton's surface, shown in **Figure 15.4**.

7. Describe the two different regions of Triton.

FIGURE 15.4

8. What is the evidence that Triton's surface has been reprocessed?

FIGURE 15.5

The image of Triton in **Figure 15.5** shows a plain of ice. The plain was probably formed by eruptions of water or water-ammonia slurry. It seems to fill the remains of ancient impact basins.

9. What region(s) on our Moon are similar in origin?

10. What is one major difference between the formation of this plain and those similar features on the Moon?

ACTIVITY 16

Classifying Meteorites

Learning Goals

In this activity you will learn about different types of meteorites and their origins. As you work on the activity you will:

1. Examine images of samples of meteorites and note dominant features and possible classification.

2. Distinguish among iron, stony-iron, and stony meteorites.

3. Classify "unknown" meteorites using supportive explanations.

4. Compare spectra of an asteroid and a meteorite and evaluate the evidence for the origin of the meteorite.

Step 1—Studying Meteorite Images

Table 16.1 shows six examples of four kinds of meteorites. Examine each image carefully. Pay particular attention to the color, fusion crust (if any), chondrules (if any), and metallic characteristics (if any). Fill in the table with as much detail as you can. You should be as specific as possible. Next, you will use your newly found knowledge in **Table 16.2** to classify some "unidentified" meteorites.

● TABLE 16.1

Six examples of four of the main types of meteorites. Refer to the color images provided by your instructor, if possible.

IDENTIFICATION AND LOCATION OF FIND	IMAGE	SPECIFIC CHARACTERISTICS NOTED		
		DOMINANT FEATURES	CHONDRULES? (DESCRIBE)	METALLIC? (DESCRIBE)
Carbonaceous chondrite Pueblito de Allende, Chihuahua State, Mexico				

(continued)

O TABLE 16.1

Continued.

IDENTIFICATION AND LOCATION OF FIND	IMAGE	SPECIFIC CHARACTERISTICS NOTED		
		DOMINANT FEATURES	CHONDRULES? (DESCRIBE)	METALLIC? (DESCRIBE)
Chondrite Cocklebiddy, Western Australia				
Iron Wolfe Creek, Mataranka, Northern Territory, Australia				
Iron Great Namaqualand, Namibia, Africa				
Stony-iron Brenham meteorite, Haviland, Kansas				
Achondrite (Eucrite) Millbillillie, Western Australia				

Step 2—Classifying Meteorites

○ TABLE 16.2

"Unidentified" meteorites for classification.

IMAGE	TYPE	REASON FOR CLASSIFICATION
(Hint: very black, with chondrules)		

Step 3— Drawing Conclusions

1. The classification of "stony" that we use includes chondrites (carbonaceous chondrites, a special kind) and achondrites (without chondrules). The stony meteorites you examined in this activity probably came from the surface of a differentiated asteroid. What does "differentiated" mean?

2. Why are smaller, rocky bodies in the Solar System (say, 1,000 meters in diameter or less) generally not differentiated while larger bodies are?

3. Let's assume that the irons, stony-irons, and stonys you examined all came from the same asteroid. How can that be? Give a possible scenario for what these meteorites have gone through over the past 4 billion or so years.

4. Explain why we believe iron meteorites must come from large asteroids.

One reliable piece of evidence about the origin of many of the meteorites suggests that at least some originated from collisions in the asteroid belt. That evidence comes from many observations from a number of widely spaced locations on Earth of a meteor falling through Earth's atmosphere. We can then use those observations to determine the meteorite's origin.

Another piece of evidence comes from comparing reflectance spectra of asteroids and meteorites. Reflectance spectroscopy is similar to stellar spectroscopy, only in this case we are studying objects that reflect the Sun's light rather than produce their own light.

Here is the evidence; you will be the judge. The graph in **Figure 16.1a** compares the reflectance spectrum of an asteroid from the asteroid belt (dots) and the reflectance spectrum of some of the small grains of the Millbillillie achondrite meteorite (solid line), shown in **Figure 16.1b.**

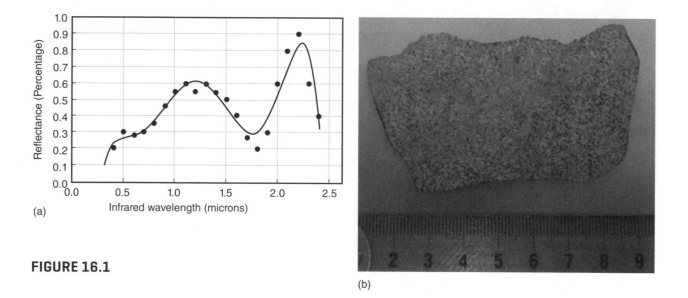

(a)

(b)

FIGURE 16.1

5. If the chemical composition of the asteroid and the meteorite are the same, then the reflectance spectra should be very similar through the entire wavelength region observed for both objects. In your opinion, is the spectral evidence solid enough to state that this meteorite was once part of the asteroid that was observed? Support your answer.

● ACTIVITY 17

Spectral Classification of Stars

Learning Goals

In this activity you will learn to determine fundamental properties of stars from a study of their spectra. By the end of this activity you will be able to:

1. Apply Wien's law to real spectra and estimate the surface temperatures of three stars.

2. Classify stars by their surface temperatures.

Step 1—Examining Spectra

On the pages that follow, you will find spectra of six stars. For three of these stars, the peak wavelength, surface temperature, and spectral type have been found for you.

1. In **Box 17.1**, study the comments on the first three stars, and then study the spectra of the last three stars. Comment on your observations for each star in the space provided at the bottom.

Step 2—Finding the Temperature

Recall that Wien's law states that the peak wavelength of blackbody radiation, λ_{peak}, is inversely proportional to its temperature, T, in kelvin. Here, we will use the unit angstrom (Å) for the wavelength because that is the unit used in the Sloan Digital Sky Survey (SDSS) spectra that we analyze here (1 Å = 10^{-10} meter):

$$\lambda_{peak} = \frac{2.9 \times 10^7}{T} \text{Å}$$

2. Solve for T, and show the equation here: _____

3. For the stars with missing information in **Table 17.1** (stars II, III, and V), visualize and then draw on the relevant spectrum of Box 17.1 a smooth curve that entirely fits over it. Figures 17.1, 17.2, and 17.3 illustrate how it is done. Find the wavelength at which each star's spectrum peaks and enter it in Table 17.1. You will fill in the other columns as this activity progresses.

4. Using the relationship found above, find the temperatures of the three unclassified stars. Enter your values in Table 17.1 in the correct temperature order, using the examples given for stars I, IV, and VI as guides. Include similar comments about peak wavelengths and temperatures for stars II, III, and V in the column adjacent to the spectra.

⊙ TABLE 17.1

Characteristics and data for six SDSS stars based on examination of their spectra.

STAR ID	PEAK WAVELENGTH (Å)	SURFACE TEMPERATURE (K)	SPECTRAL TYPE
I	<3,000 Å	>10,000 K	B
II			
III			
IV	~4,800 Å	~6000 K	F or G
V			
VI	>9,200 Å	<3100 K	M

Step 3—Finding the Spectral Type

5. Use the information in **Table 17.2** to find the spectral type of stars II, III, and V. Enter the type of each star into Table 17.1.

6. Generally describe how the spectra change in appearance from the hottest star in the sample, star I, down to the coolest star, star VI. Address how the overall shapes of the spectra look, how the number and deepness of the absorption lines change, and at least one other contrasting feature you note.

⊙ TABLE 17.2

The spectral types associated with each surface temperature range for stars.

SPECTRAL TYPE	TEMPERATURE RANGE (K)
O	>33,000
B	10,000–33,000
A	7500–10,000
F	6000–7500
G	5200–6000
K	3700–5200
M	<3700

7. All stars have essentially the same composition; for these stars, only the temperatures are different. Summarize the physical reasons why the spectra differ for the hot stars versus the cool stars in your classification. That is, explain what is going on with the electrons in the atoms that are in the atmospheres of these stars.

⊙ BOX 17.1

A small sample of Sloan Digital Sky Survey stars for determining surface temperatures and classifications.

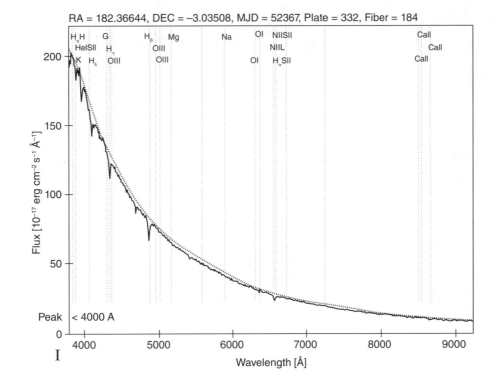

RA = 182.36644, DEC = −3.03508, MJD = 52367, Plate = 332, Fiber = 184

FIGURE 17.1

Star I is the hottest star in our sample. Its peak wavelength lies at a much shorter wavelength than 3,000 Å.

Calculating its surface temperature at 3,000 Å and assuming it is hotter than that result, we find: $T \geq 10,000$ K.

Its spectral type is B according to Table 17.2.

(continued)

⊙ BOX 17.1

Continued.

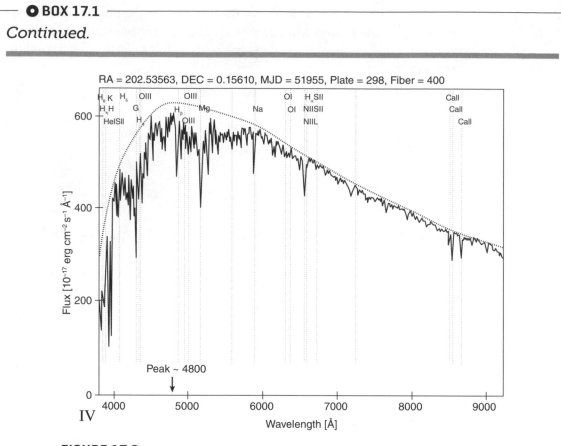

FIGURE 17.2

Star IV has a peak wavelength estimated at about 4,800 Å. Notice that we have drawn a smooth curve over the shape of the spectrum, remembering what a blackbody or thermal radiation curve looks like in general.

This peak wavelength gives us a surface temperature of ~6000 K.

The spectral type is F or G.

○ **BOX 17.1**

Continued.

RA = 179.18272, DEC = −1.14384, MJD = 51930, Plate = 285, Fiber = 242

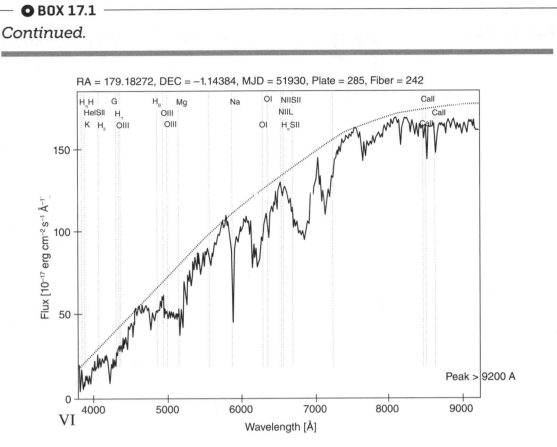

FIGURE 17.3

Star VI is the coolest star in our sample. It shows lots of absorption lines and some deep absorption features that span a range of wavelengths. It is the deep absorption features that indicate molecules are present in its atmosphere. We probably do not see its actual peak wavelength in this spectrum and so assume it is at a wavelength longer than 9,200 Å.

If the peak were 9,200 Å, then the surface temperature would be 3100 K.

Its spectral type is M.

(continued)

○ **BOX 17.1**

Continued.

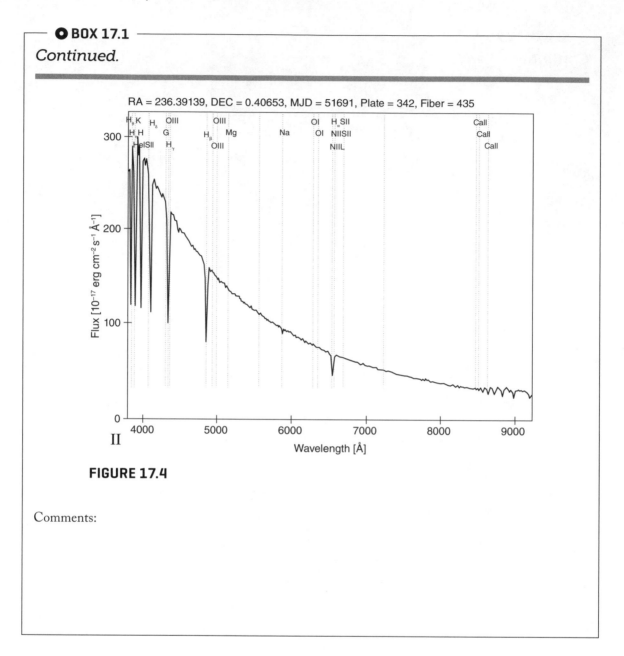

FIGURE 17.4

Comments:

● BOX 17.1

Continued.

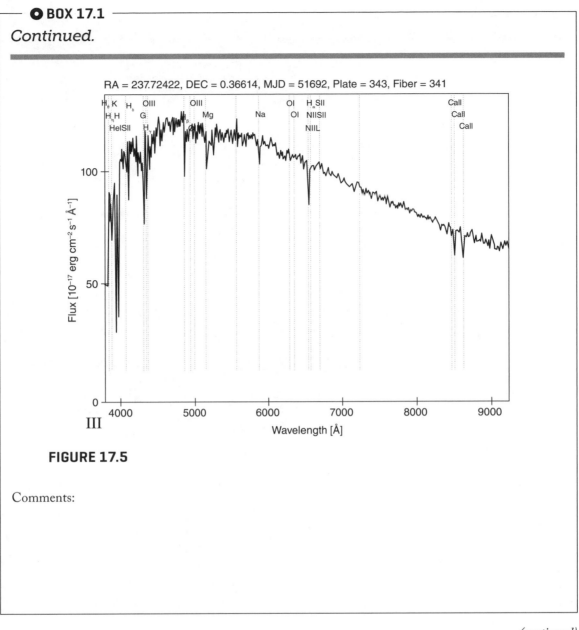

FIGURE 17.5

Comments:

(continued)

⊙ BOX 17.1

Continued.

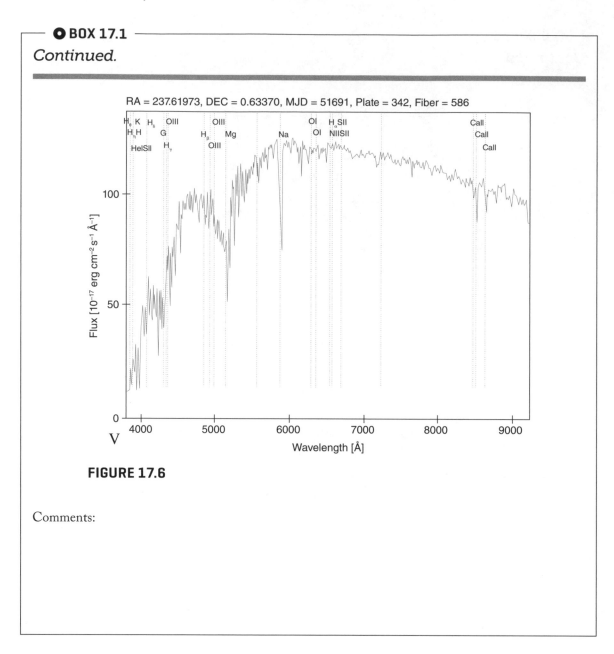

FIGURE 17.6

Comments:

ACTIVITY 18

Finding Distances to Stars Using Parallax Measurements

Learning Goals

In this activity you will determine a relationship between distance and apparent motion of a nearby object when viewed from two vantage points. You will apply this relationship to the measuring of distances to stars. As you work through this activity you will:

1. Demonstrate parallax of a nearby object, at different distances from your eyes, relative to a distant scale.

2. Derive the relationship between the distance of the nearby object and the sizes of its apparent shifts relative to a distant scale.

3. Apply this knowledge to the measured parallax angles of stars.

Step 1—Measuring a Parallax Angle

You can see the parallax effect by holding your thumb out at arm's length, looking at a distant object, and alternately opening and closing each eye. The thumb will seem to jump back and forth relative to the background. This is because the centers of our eyes are about 7 centimeters (cm) apart from each other, so each eye has a slightly different point of view.

1. First, make a prediction about the relationship between the distance of a pencil (or other long, thin object) and the number of fine grid marks across which it will appear to jump. Which relationship, of those shown in **Figure 18.1**, would you predict is correct?

_____ $y = x$

_____ $y = -x + 3$

_____ $y = 1/x$

_____ $y = \text{sqrt}(x)$

Now let's test how the parallax of an object varies with distance using the baseline measurement from our eyes. Although using a meter stick would give more quantitative results, we can get respectable data by the use of an arm and a pencil. Be sure to hold your arm straight out toward the distant scale from which

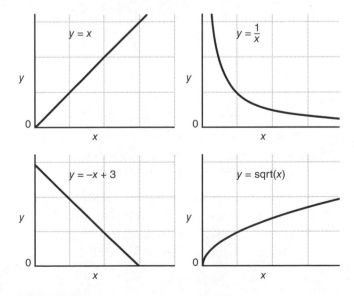

FIGURE 18.1

FIGURE 18.2

you will be determining the shift, and have the pointed end of the pencil up. **Figure 18.2** illustrates what to do.

2. Your nose and eyes should be pointing straight down your arm. Approximate the distance from your eyes to the end of your finger and enter it as the first distance in Table 1; it won't matter if it is accurate as long as you are consistent in your measurements. Be sure to note the units you will be using in your measurements.

   a. Hold your arm straight and place the pencil out as far as you can reach with your other arm. Alternate opening and closing your eyes and judge the shift of the pencil point with the distant scale. Write the number of grid marks that it shifted on the first line of Table 18.1.

○ TABLE 18.1

Data table for measurements.

DISTANCE UNITS USED: _____	NUMBER OF GRID MARKS

   b. Now move the pencil to half of the original distance and alternate opening and closing your eyes. Judge the shift of the pencil point with the distant scale and write the number of grid marks that it shifted on the second line of Table 18.1.

   c. Repeat this process for at least six more measurements, placing the pencil one-fourth the distance to the end of your finger, then three-fourths, one-third, and so on. Enter the approximations of the distances and the number of grid marks that the pencil point shifted on the remaining lines of Table 18.1.

3. Using **Figure 18.3**, graph the number of fine grid marks the pencil jumped versus the distance from your eyes. Be sure to label both axes.

4. State the relationship between the distance from the baseline of your eyes and the number of fine grid marks the pencil jumped. Show your logic here. Was your earlier prediction correct?

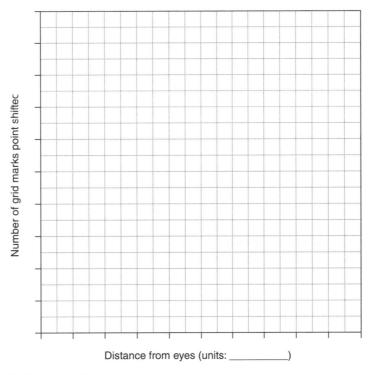

FIGURE 18.3

Let's take a look at the approximate relationship between distance and parallax from a different viewpoint by examining the angles shown in **Figure 18.4**.

5. The distance d_2 is twice the distance d_1. Does it qualitatively appear that angle a_2 is one-half of a_1? Yes or No

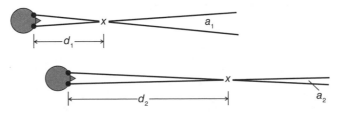

FIGURE 18.4

Step 2—Application to Stars

Astronomical parallax is measured as shown in **Figure 18.5**. *B* indicates the diameter of Earth's orbit. At right is a star at great distance, *d*. The angle, α, is a measurable quantity. When the distance is large enough that the parallax angle is very small, which is certainly true for all stars except the Sun, the parallax angle, *p*, is proportional to the inverse of the distance (1/*d*). Conversely, if we can measure the parallax angle, we know that the distance to the object is proportional to the inverse of that angle.

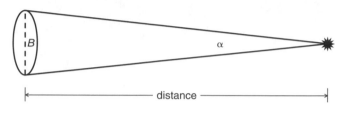

FIGURE 18.5

Note: Although the baseline for the measurement of parallax angles for stars is the diameter of Earth's orbit, *B*, we use instead the value of 1 astronomical unit (AU), the radius of Earth's orbit, and the stellar parallax angle, *p*, is half the angle, α. This gives the distance of a star from the Sun.

The longer the baseline, the more objects move relative to the background—objects can be farther away and still have a measurable parallax.

6. Which of the following scenarios would be the more desirable? _____ Explain your reasoning.

   a. a satellite orbiting the Sun at Jupiter's distance (5.2 AU)

   b. a telescope sitting on Pluto (Pluto is 40 AU from the Sun)

7. Which of the following scenarios do you think would be the more feasible? _____ Explain your reasoning.

   a. a satellite orbiting high above Earth (above the atmosphere)

   b. a telescope on the Moon

Because the stars are so far away, the distance at which we can measure a parallax is directly proportional to the length of the baseline.

$$\text{Baseline} = \alpha \times \text{distance} \quad \text{or} \quad B = \alpha \times d$$

8. Suppose astronomers put a telescope on Pluto, ~40 AU from the Sun, beaming back information to us. How much **farther** will they be able to measure accurate parallaxes compared to their work here on Earth?

9. Imagine that you measure the parallax of two stars in the constellation Leo. Regulus has one-half the parallax angle of Denebola. What do you immediately know about the relative distances of these stars from Earth?

We can now investigate more thoroughly why parsecs and arcseconds are convenient for astronomy. A star that is 1 parsec (pc) away has a parallax angle of 1 arcsecond (arcsec; see **Figure 18.6**). A star that is 2 parsecs away has a parallax angle of 0.5 arcsecond, and so on. The unit "parsec" makes converting from parallax angle to linear distance straightforward.

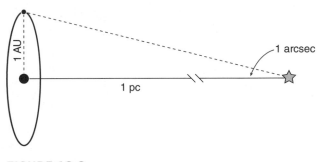

FIGURE 18.6

10. Rank the stars in **Table 18.2** in order of distance from Earth.

⊙ TABLE 18.2

Rank: closest = 1, farthest = 4.

RANK	STAR NAME	PARALLAX ANGLE (ARCSEC)
	Antares	0.024
	Ross 780	0.213
	Regulus	0.045
	Betelgeuse	0.009

11. Calculate the distances to the stars listed in **Table 18.3**. The parallax angle is given in arcseconds, and the distance will come out in parsecs.

○ TABLE 18.3

Calculating distances to stars.

STAR	PARALLAX ANGLE (ARCSECONDS)	DISTANCE (PARSECS)
Arcturus	0.090	
Procyon	0.288	
Hadar	0.006	
Altair	0.194	

12. Briefly explain how we calculate the distance to a star in parsecs by measuring its parallax angle in arcseconds. Give an example. List an advantage and a disadvantage of this method.

ACTIVITY 19

Analyzing a Coronal Mass Ejection from the Sun

Learning Goals

In this activity, you will take a close look at a coronal mass ejection (CME) from the Sun, calculate the speed of a clump of material, and determine how much notice scientists have before the charged particles reach Earth. After completing the activity you will be able to:

1. Calculate the speed of the material ejected.

2. Determine how much advanced warning we would have.

3. Consider the source(s) of errors and uncertainties.

4. Summarize why it is important for us to predict CMEs accurately.

Step 1—Finding the Speed of a CME

1. Examine the images in **Figure 19.1**, and identify the "clump" of material that can be traced over a series of images. The ruler that has been overlaid should help.

2. Find the center of the clump in each image, and measure how far it has traveled away from the Sun. Insert your data into **Table 19.1**.

● TABLE 19.1

Measurements of the distance traveled by the coronal mass ejection from the Sun.

TIME (UT)	HOURS	SCALED DISTANCE ON IMAGE (MILLIMETERS)
10:45	0	
11:42	1	
12:42	2	
13:42	3	
14:42	4	
15:42	5	

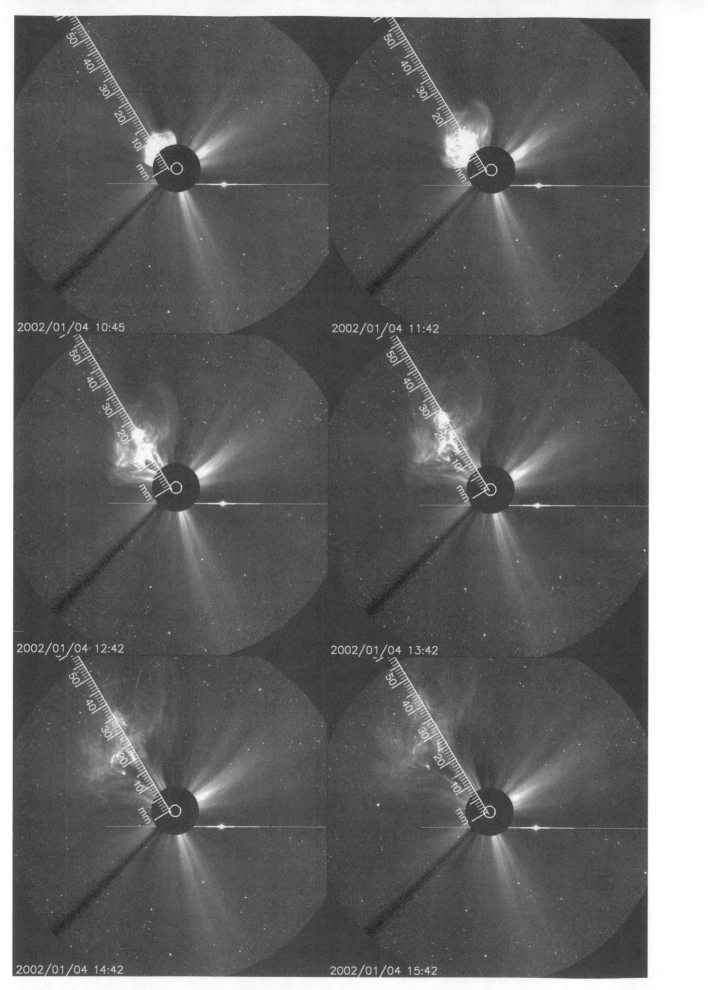

FIGURE 19.1

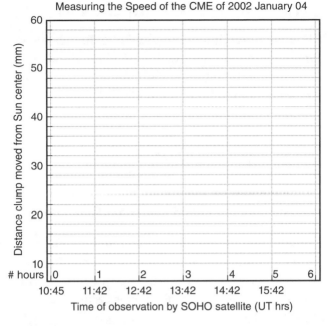

FIGURE 19.2

3. Use **Figure 19.2** to graph your results. Make a best-fit line through these data by drawing a straight line that follows the trend in the data.

4. Find the slope of the line by picking two points and subtracting the y-position of the first point from the y-position of the second point. Divide this difference by the difference between the x-positions, found by subtracting the x-position of the first point from the x-position of the second point. Your answer will come out in millimeters per hour (mm/h).

What is the slope of the data in millimeters per hour ?

_____

5. Calculate the scaling factor for the images: The small central circle shown in white represents the linear size of the Sun as represented by the image. The actual diameter of the Sun is approximately 1.4×10^6 kilometers (km).

   a. Measure the diameter of the Sun in the image in millimeters.

Diameter of the Sun in millimeters:

_____

Divide the actual diameter of the Sun by this value. For example, if the small central circle measured 7 mm across, then in this image, every 7 mm represents 1.4×10^6 km per 7 mm, or

$$\frac{1.4 \times 10^6 \text{ km}}{7 \text{ mm}} = \frac{2 \times 10^5 \text{ km}}{1 \text{ mm}} = 2 \times 10^5 \text{ km/mm}$$

   b. What is the scale of the images? _____ km/mm

6. Multiply your slope (question 4) by the image scale to find the average speed of the CME over the 5 hours of observation.

What is the slope of the data in kilometers per hour (km/h)?

_____

Step 2—Finding the Time It Takes for the CME to Arrive

7. This particular CME was not traveling toward Earth. Others, however, occasionally do. The distance between Earth and the Sun is about 1.5×10^8 km. Dividing this distance by the speed of the CME will give the time it takes the CME to arrive at Earth; that is, how much advance notice scientists could have. Consider a similar CME directed at Earth.

    How many hours would it take a similar CME to reach Earth? _____

    How many days would it take a similar CME to reach Earth? _____

Step 3—Considering Uncertainties and Summarizing

8. Your results depend primarily on an accurate measurement of the Sun's diameter, locating the same clump in each of the images, and accurately determining the center of the clump as the material spreads out. Let's consider just the measurement of the Sun's diameter on the image, as that error is systematic—our measurements may be too large or too small, and either way will affect our results.

    a. What is your original measurement for the diameter of the Sun? _____ mm

    b. What would the range of diameters be if your measurement were 25 percent off? (Example: For the 7-mm diameter given earlier, there would be an uncertainty of 7 mm ± 25 percent = 7 mm ± 1.8 mm [rounded off]. The largest diameter would be 8.8 mm; the smallest 5.2 mm. The scale factor would then have a range of 269,000 km/mm for the largest value and 160,000 km/mm for the smallest value.)

        i. Smallest: _____

        ii. Largest: _____

    c. What would the range of scale factors be if your measurement were 25 percent off? (Continuing the example from above: If the largest reasonable diameter is 8.8 mm and the smallest is 5.2 mm, then the scale factor would have a range of 269,000 km/mm for the largest value and 160,000 km/mm for the smallest value.)

        i. Smallest: _____

        ii. Largest: _____

    d. Continue through the calculations based on the steps above. What would be the range of speeds in kilometers per hour?

        i. Smallest: _____

        ii. Largest: _____

9. Summarize why it is important that we find a way to calculate the amount of advance notice we have before the material from a violent CME reaches Earth. Why are time and money being spent figuring out a way to predict solar flares that create CMEs?

ACTIVITY 20

The Stuff Between the Stars

Learning Goals

In this activity you will study images of various locations in the interstellar medium to learn to distinguish among the reasons an object might look red. You will also learn to:

1. Distinguish among the red emission radiation created in H II regions, interstellar reddening (de-bluing) caused by dust, and dust clouds emitting infrared light due to thermal radiation.

2. Explain how astronomers help us to "see" objects and features that give off light that is outside of the visible part of the spectrum.

Step 1—Mapping the North America Nebula

Your instructor will show you a pair of images of the North America Nebula for comparison: one is an image of the nebula taken in visible light, and the other is an image of the nebula taken in the infrared by the Spitzer Space Telescope.

1. In **Figure 20.1**, make a map of the North America Nebula imaged at **visible** wavelengths, filling in just enough of the details so that you can distinguish the emission nebula (reddish in the image provided by your instructor) from the dark nebula (dark gray to black in the image provided by your instructor). Locate a few of the brighter stars. We have added the Pelican Nebula seen at the right in Figure 20.1 as an example of mapping.

FIGURE 20.1

2. List two significant ways that the visible-light image differs from the infrared-light image.

3. What mechanisms, events, or processes are producing the light in each case?

4. The image taken in visible light includes wavelengths between _____ nm and _____ nm (approximately).

5. The wavelength range for the red part of the spectrum is roughly between _____ nm and _____ nm.

6. The strongest emission line seen at red wavelengths comes from excited hydrogen atoms at a wavelength of _____ nm when the electrons jump down from energy level _____ to _____ and emit photons.

Step 2—Discovering Why Interstellar Reddening of Light Occurs

Figure 20.2 will guide you to a better understanding of the reddening of starlight due to interstellar dust. Figure 20.2a is an image of the top region of the Horsehead Nebula in Orion. Figure 20.2b shows what the Horsehead would look like if you could view it from the side; that is, we have to imagine "turning" the image counterclockwise and looking at it from that perspective. The light from the stars is being "de-blued"; that is, the light being emitted from the stars at blue wavelengths is scattered away from our line of sight, leaving the longer red wavelengths—at the visible part of the spectrum—to be seen. This redness results from an entirely different process than that of a star whose blackbody radiation peaks at red wavelengths.

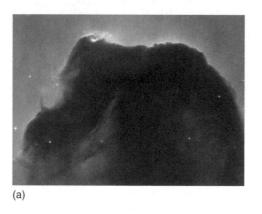

(a)

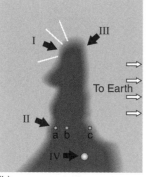

(b)

FIGURE 20.2

7. Which stars in the actual color image provided by your instructor seem to be "reddened" by the dust of the Horsehead Nebula? Circle those stars in Figure 20.2a.

8. In Figure 20.2b, where would a likely location be for those stars whose light has been reddened (locations are labeled, a, b, and c in the figure)?

    a. behind the dust column

    b. deep inside the dust column

    c. in front of the dust column

    d. Either a or b are possible locations for reddening to occur.

9. Indicate on Figure 20.2b where the scattered blue light would go that caused the star's light to be reddened.

10. Is this reddened light coming from the stars seen at visible or infrared wavelengths? Explain.

11. Referring to the information about Figure 20.2b summarized earlier in this activity and using the sketched "side view" of the figure, where would you locate the "young star still embedded in its nursery of gas and dust"? _____ I, _____ II, _____ III, or _____ IV?

Step 3—Viewing the Interstellar Medium at Infrared Wavelengths

Your instructor will show you an image titled "An Audience Favorite Nebula," from the Spitzer Space Telescope. This image was part of a program to identify interstellar bubbles.

12. The image of this region of space displays the colors blue, green, red, and yellow. Does this mean that the nebula is emitting light in these colors? Explain.

13. For the wavelengths colored red, do they indicate reddening of visible light by the dust or rather the dust glowing because it is warm? Support your answer.

Step 4—Infrared Light from Warm Dust

Your instructor will show you an image from the Spitzer Space Telescope titled "Stars Brewing in Cygnus X." This image was chosen because it encompasses an angular size of the night sky that is around eight times the size of the full Moon, yet is invisible to the naked eye.

14. From the Spitzer image, which color represents a mapping of the warmest dust?

15. What is the temperature of that dust?

16. Which color represents a mapping of the coolest dust?

17. What is the temperature of that dust?

18. Do the red-colored parts of this region indicate reddening of starlight or dust glowing? Explain.

19. Contrast in detail how the red-colored glow in this image differs from the red emission observed in the North America Nebula.

ACTIVITY 21

Investigating the Crab Nebula and Pulsar

Learning Goals

In this activity you will investigate a supernova remnant, finding the speed at which a feature embedded in this explosion is traveling outward. You will also:

1. Confirm which images had to be taken from a space-based observatory.

2. Hypothesize as to why the images taken at different wavelengths have different angular sizes (images have the same scale).

3. Calculate the speed of a distinctive "wave" from the pulsar and compare it to the speed of light.

Step 1—The Crab Nebula

Figure 21.1 shows the Crab Nebula in four wavelength regions. Consider these images, and answer the following questions.

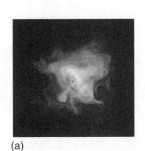
(a)

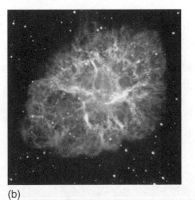
(b)

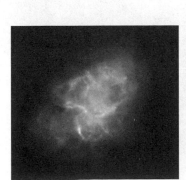
(c)

(d)

FIGURE 21.1

1. Think about the regions of the electromagnetic spectrum where light reaches Earth's surface and other regions where the atmosphere blocks the light. Ignoring any sharpness or clarity of the images, which two of the images *had* to be taken either from a high mountain or a space telescope?

2. Which type of radiation is revealing events with the highest energy?

3. Which type of radiation represents events having the lowest energy?

4. The size of the radio emitting part of the nebula is about 3.5 times that of the X-ray emitting part. Based on your knowledge of synchrotron radiation and how electrons lose energy to radiation, does this difference in extent make sense? Explain.

Step 2—The Crab Pulsar

5. All pulsars are neutron stars, but not all neutron stars are pulsars. From our perspective, what makes a neutron star a pulsar?

Recent observations of the Crab Nebula have detected rapidly changing wisps of material. These not-so-subtle changes are shown in the Hubble Space Telescope images shown in **Figure 21.2**. These include wisplike structures that move outward away from the pulsar at almost half the

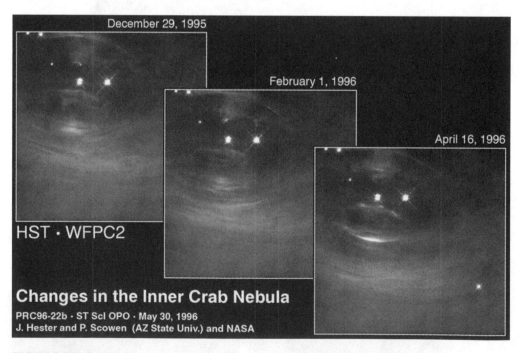

FIGURE 21.2

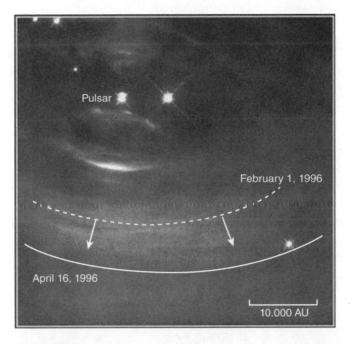

Pulsar

February 1, 1996

April 16, 1996

10.000 AU

FIGURE 21.3

speed of light, as well as a mysterious "halo" that remains stationary, but grows brighter then fainter over time. **Figure 21.3** "zooms in" to the core of the nebula: the region around the pulsar. Based on the original image scale, it appears that the white-line "wave" moved about 6,000 astronomical units (AU) between February 1 and April 16, 1996 (75 days).

6. What is the approximate speed for this "wave" in kilometers per second (km/s)? You need the following data to calculate the answer to this question: 1 AU = 150,000,000 km; 1 day = 60 × 60 × 24 = 86,400 seconds.

7. At what fraction of the speed of light, 300,000 km/s, was this wave moving? (Show the equation and the numbers you used.)

8. Consider your calculation of the speed of the material. Is this a reasonable speed for the material to travel? Could it be a shockwave that is fooling us? Explain your thoughts.

9. Let's put all of this into perspective with the following comparison: The diameter of a neutron star is given as approximately 20 km. How much of your city would the neutron star cover if it were to sit on a pedestal, say at city hall? Bring in surrounding neighborhoods if necessary.

ACTIVITY 22

Determining Ages of Star Clusters

Learning Goals

In this activity you will find the ages of star clusters (and the stars in them) by considering the rate at which stars fuse hydrogen to helium and by studying the color-magnitude diagrams of clusters. After completing the activity, you will be able to:

1. State the relationship between the color of a main-sequence star and its main-sequence lifetime.

2. List the steps for determining the age of a cluster given the cluster's "turnoff" B – V color.

3. Determine cluster ages using their color-magnitude diagrams.

4. Summarize the complete process of how astronomers estimate the ages of star clusters.

Step 1—The Meaning of the B – V Color Index

Hot stars are bluer than cool stars. Cool stars appear redder than hot stars. The "B – V color index" (often shortened to just "B – V") is a way of quantifying this using two different filters. **Figure 22.1** compares blackbody curves for objects at two different temperatures and filter wavelengths for B (blue) and V (visual; yellow-green) filters. The hottest stars have B – V color indices

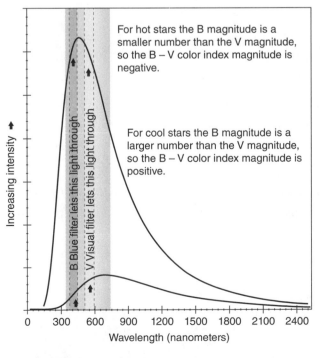

FIGURE 22.1

close to −0.5 or less, while the coolest stars have B − V color indices close to 2.0. Other stars are somewhere in between. To find the B − V color index, astronomers first measure the apparent magnitude with two different filters: B and V. They then subtract the V magnitude from the B magnitude to find the color index.

As you know from experience, it is possible for a car to hold more fuel but run out of that fuel faster than a car that holds less fuel. This idea applies directly to stars. Stars spend most of their lives on the main sequence where they convert the hydrogen in their cores into helium and energy. Stars that are more massive and have more hydrogen in their cores will always run out of hydrogen fuel sooner than those that are less massive. Put another way: The more massive a star is, the shorter its life span.

1. Calculate the B − V values for the following main-sequence stars and fill in **Table 22.1**. Notice that the B − V color index tells us the difference in the magnitudes as measured through each filter.

● TABLE 22.1

B and V color indices for three main-sequence stars.

STAR	B-FILTER MAGNITUDE	V-FILTER MAGNITUDE	B − V COLOR INDEX
Barnard's Star M4 V	11.24	9.51	
Mintaka O9 V	2.01	2.41	
Zavijava F9 V	4.16	3.61	

---

2. For stars on the main sequence, the spectral type is correlated with luminosity. Which of these three stars will spend the shortest amount of time on the main sequence?

3. Which will spend the longest amount of time on the main sequence?

4. Which stars are brighter in the visual filter than they are in the blue one?

5. The answer to the previous question implies that there might be a relationship between the color index of a star and its main-sequence lifetime. What is that possible relationship?

Step 2—The quantitative relationship between the Color Index and the Lifetime of a Star on the Main Sequence

So far, we have been somewhat *qualitative* in our descriptions: Main-sequence stars with a smaller B – V value will spend less time on the main sequence because as we go to smaller B – V values, we are moving up the main sequence to the region where the hottest, most luminous, massive stars lie. There is a *quantitative* relationship between these two quantities (see **Table 22.2**) that we now consider.

⊙ TABLE 22.2

Comparison of spectral types of stars, their masses, color indices, and main-sequence lifetimes.

PARAMETER	SPECTRAL TYPE						
	O5	B0	A0	F0	G0	K0	M0
Mass (solar)	40	15	3.5	1.7	1.1	0.8	0.5
B – V color	**−1.2**	**−0.3**	**0.0**	**0.3**	**0.6**	**0.8**	**1.4**
Main-sequence lifetime (years)	1.0×10^6	1.1×10^7	4.4×10^8	3.0×10^9	8.0×10^9	1.7×10^{10}	5.6×10^{10}

Figure 22.2 contains overlapping color-magnitude diagrams for two different star clusters: NGC 2362 and M68. On the figure, note the numbered star symbols for NGC 2362 stars and numbered solid circles for M68 stars.

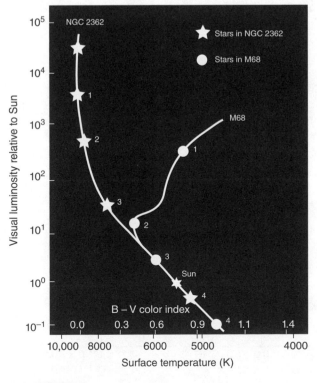

FIGURE 22.2

The turnoff for NGC 2362 appears to be at a B – V value of 0.0. This means the cluster is about 4.4×10^8 years old. Stars in this cluster are marked with a star symbol.

6. How old is star 1? _____

7. How old is star 2? _____

8. How old is star 4? _____

Stars in the cluster M68 are marked with a solid circle. Star (solid-circle) 2 is in M68. It is turning off the main sequence at a B – V of about 0.4, indicating an age of about 5×10^9 years.

9. About how old is the cluster M68? _____

10. About how old is star (solid-circle) 1? _____

11. About how old is star (solid-circle) 3? _____

12. State a "rule" about figuring out the age of a star cluster based on the value of the B – V color index of the turnoff point from the main sequence.

Step 3—Applying the Rule to Observations of Real Star Clusters

Imagine that we observed four different clusters in our galaxy. **Figure 22.3** shows the results of these observations after we plotted the V magnitude and the B – V color index.

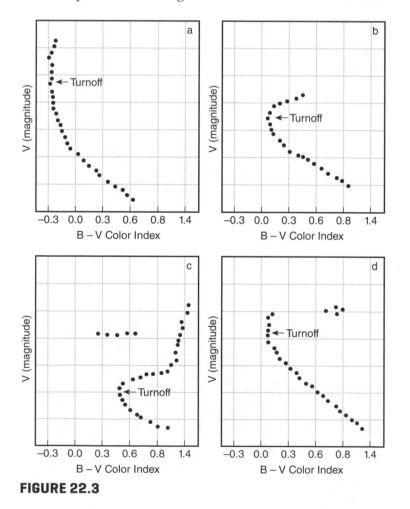

FIGURE 22.3

13. Rank the ages of the different clusters shown, from youngest to oldest.

Youngest cluster _____ _____ _____ _____ Oldest cluster

Explain the logic you used.

Alternatively, imagine that these are theoretical color-magnitude diagrams of the stars in a single cluster, evolving over time.

14. Rank the plots in the order in which they would occur.

Earliest _____ _____ _____ _____ Latest

Explain the logic you used.

Determine the B – V value of the turnoff (TO) and the age of each of the clusters in **Figure 22.4**. The y-axis is the measured V apparent magnitude. (These plots represent real data, and may contain stars that are in the images, but not part of the cluster. Thus, they may not be at the same distance as the cluster.)

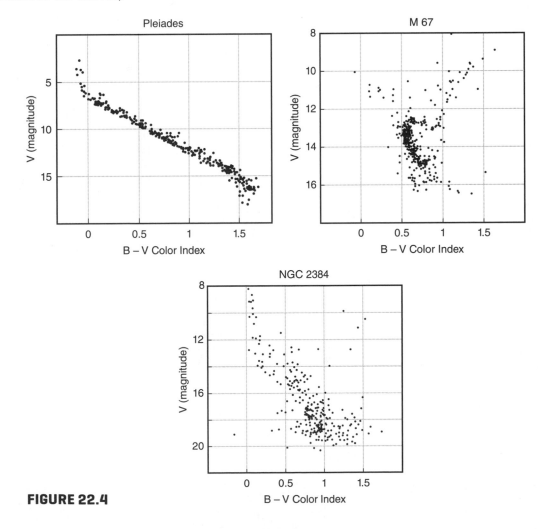

FIGURE 22.4

15. Pleiades:

a. B – V turnoff value: _____

b. Age: _____

16. M67:

a. B – V turnoff value: _____

b. Age: _____

17. NGC 2384:

a. B – V turnoff value: _____

b. Age: _____

18. Explain why our placing observations of stars on a color-magnitude diagram allows us to determine the age of a cluster. This should *not* be a step-by-step how-to; rather, your response should incorporate ideas such as distance, luminosity, and lifetimes of stars, how clusters are formed, and similar concepts.

ACTIVITY 23

Bent Space and Black Holes

Learning Goals

In this activity you will use thought experiments and the metaphor of a rubber sheet to explore the interaction of black holes with objects that pass near them. As you work through this activity you will also:

1. Explain the concept of bent space.

2. Demonstrate an understanding of how the speed of an object affects its interaction with a massive body.

3. Define *event horizon*.

4. Extend the concept of an event horizon to objects that travel at speeds slower than light.

5. Extrapolate the two-dimensional concepts to three dimensions.

Step 1—Considering Black Holes in Two Dimensions

Einstein's idea of objects bending or warping space applies to anything having mass—the more massive an object is, the more space is bent. Here we will consider the ultimate in bent space, that caused by black holes. Because we can't actually grab a black hole and bring it into the lab and because scientists have never actually observed one directly, we can conduct only "thought experiments" to explore their properties.

1. Imagine a big rubber sheet, as shown in **Figure 23.1**. It is very stiff and not easily stretched, but it does have some "give" to it. Imagine that you are holding the sheet parallel to the ground and roll some golf balls across it in directions you choose. Sketch their paths across the sheet on the figure.

2. Now imagine that we put a bowling ball (much heavier than a golf ball) in the middle of the sheet so that it makes a big, slope-sided pit, as shown in **Figure 23.2a**. (Figure 23.2b is provided to aid in your visualization of this task.) Roll three more golf ball across the sheet according to the list given next and sketch their paths on the figure.

   a. The golf ball is far from the bowling ball, near the edge of the sheet.

   b. The golf ball goes directly toward the bowling ball.

   c. The golf ball comes close to the bowling ball but not directly at it.

3. Think again. In each of the three cases of question 2, how would things change if the golf balls were moving very quickly?

FIGURE 23.1

4. In each of the three cases of question 2, how would things change if the golf balls were moving very slowly?

5. Imagine that you send lots of golf balls into the pit. What happens to the depth and width of the pit as the golf balls fall into the center near the bowling ball?

6. So far you have been working with two examples of space. The first example has very little bend to the space; in the second example, a more massive object placed in the center bends space quite a bit. How do these examples explain the concept of bent space in terms of the objects rolling across the rubber sheet?

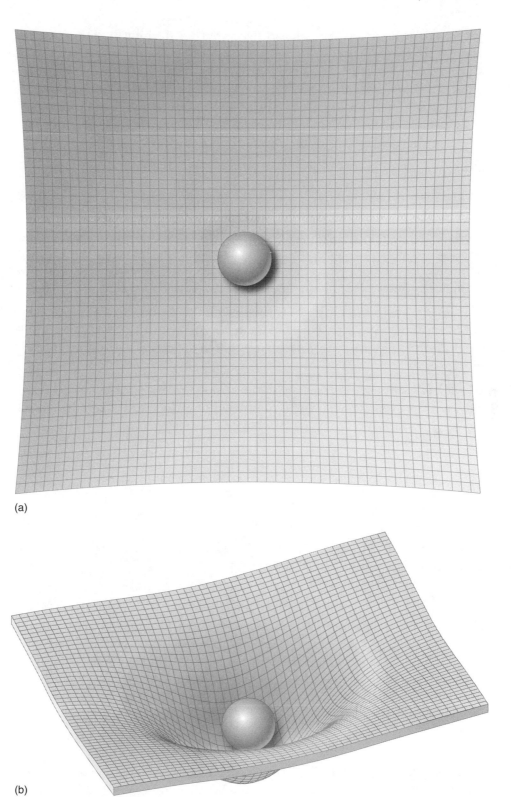

(a)

(b)

FIGURE 23.2

7. The pit is a fair analogy for a black hole. Objects outside the pit will "know" that the black hole is there because the sheet is sloping, but they won't be captured unless they come within the *event horizon*. This is the location at which light bends exactly into a circle around the black hole, as shown in **Figure 23.3a**. (Figure 23.3b is provided to help in your visualization of this task.) Think about light as though it were grains of sand rolling across the sheet extremely fast. Sketch the path of one grain-of-sand light particle if it were to travel

   a. far from the bowling ball, near the edge of the sheet.

   b. directly toward the bowling ball.

   c. on a line that is tangent to (just touches) the event horizon.

   d. inside the event horizon, but not directly at the bowling ball.

Stars, people, planets, all things interact in this way because of gravity. In the case of black holes it is a little more accurate to think of the bowling ball as making a very, very deep hole (so nothing can bounce off, for example), but the sheet is still bent in the same way.

8. Now think again about what happens when you roll the slower-than-light golf balls past the pit. Sketch and label a circle on Figure 23.3a that would mark a "golf ball event horizon." Consider the case where the golf balls come closer to the center than this circle and they cannot escape.

9. Is the "golf ball event horizon" that you sketched closer or farther from the bowling ball than the light event horizon? Why? (Hint: Think about how the speed affects the motion.)

10. Suppose that you roll another bowling ball across the sheet. What happens to the sheet when the second bowling ball falls in after the first?

11. How would this affect the paths of the golf balls? How about the example of a grain-of-sand light particle?

12. From your experience with these thought experiments, give a definition of *event horizon* that is consistent both with golf balls rolling on a rubber sheet that is bent and with light interacting with the bending of space around a black hole.

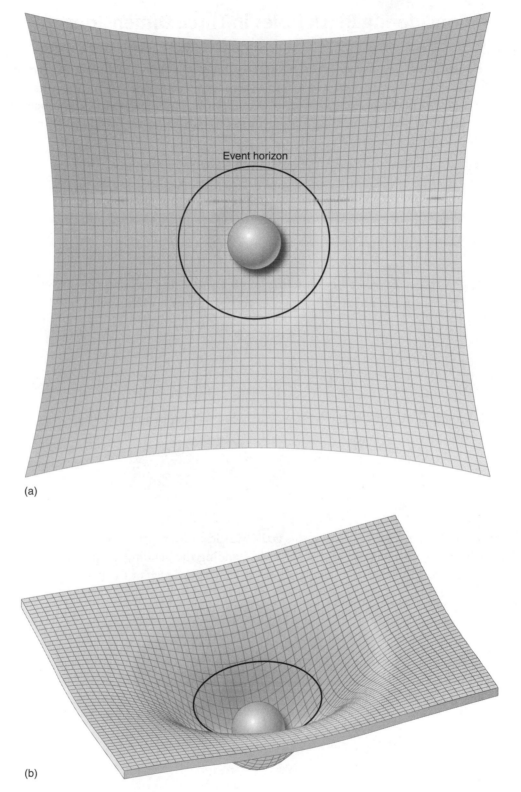

(a)

(b)

FIGURE 23.3

Step 2—Considering Black Holes in Three Dimensions

The thought experiments that you have just carried out are two-dimensional metaphors for a three-dimensional bending of space and time (although we don't consider time here). To think about black holes properly, you need to add a third dimension in which the same effects apply. This is much more difficult than it might seem because now the rubber sheet is no longer flat with a pit in the center, but is bent from all directions toward the center no matter from which direction golf balls or light approaches.

13. Imagine that a golf ball is fired into Figure 23.3a from above the page.

    a. If the golf ball starts out aimed at the very edge of the sheet, what will happen to it? Why?

    b. If the golf ball is aimed at a point to the left of the bowling ball and falls in between the edge of the sheet but not too close to the "golf ball event horizon," what will happen to it? Why?

    c. If the golf ball is aimed at the bowling ball, what will happen to it? Why?

14. Summarize the concepts covered in this activity. Add to your summary a few of the ways that the rubber-sheet analogy works for understanding the bending of space and a few of the ways it fails. How has this activity added to your knowledge of the behavior of objects approaching black holes?

ACTIVITY 24

Light Travel Time and the Size of a Quasar

Learning Goals

Astronomers use variability in the emission from active galaxies to determine the size of the emitting region and the total energy emitted. In this activity, you will connect these observations to the theory that active galaxies are powered by supermassive black holes. You will also learn to:

1. Determine the size of the emitting region of a "new" quasar based on a brief intense flare.

2. Calculate the energy released by the flare and compare that to the energy of the Sun.

3. Evaluate the evidence as to whether it supports the presence of a supermassive black hole.

Step 1—Measuring the Size of an Active Galaxy Flare Region

The brightness of Active Galactic Nuclei (AGN) can change dramatically over the span of months, weeks, days, and even minutes. Astronomers can find the size of the emitting region of an active galaxy from the speed of light and limits on how quickly an object can change its brightness.

Imagine an object 1 light-week in diameter, as sketched in **Figure 24.1**. Suppose that the entire object emits a brief (~1 second) flash of light. Light from the part of the object nearest to Earth will arrive here first. Light from the far side of the object will arrive 1 week later, because this light not only must cross the distance to Earth, but it also must cross the object as well. Because of this delay in the arrival time of the light from the back of the object, we will observe a gradual change in brightness that lasts a full week. The *rise time* is the time for the brightness to increase to its maximum. The *decay time* is the time for the brightness to drop back down to the "normal" level. Similarly, an object that is 1 light-year in diameter will take 1 year for the brightness to vary, as shown in **Figure 24.2**.

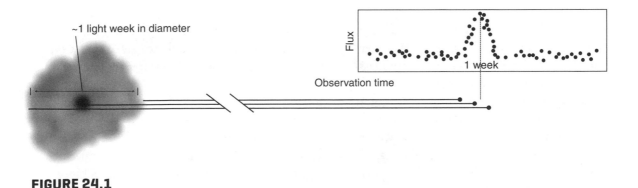

FIGURE 24.1

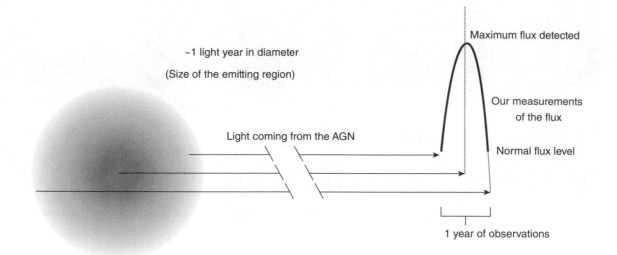

FIGURE 24.2

The size of the emitting region, d, is found by multiplying the speed of light, c, by the variation time, Δt:

$$d = c \times \Delta t$$

This variation time must be in seconds, so for the example of an AGN with a 1-week variation time, we must first convert 1 week (7 days) into seconds:

$$\Delta t = 7 \text{ d} \times 24 \text{ h/d} \times 60 \text{ min/h} \times 60 \text{ s/min}$$

$$= 604{,}800 \text{ s}$$

Then we can multiply this time by the speed of light to find the diameter:

$$d = c \times \Delta t$$

$$= (3 \times 10^8 \text{ m/s}) \times (6.04 \times 10^5 \text{ s})$$

$$= 1.81 \times 10^{14} \text{ m}$$

$$= 18.1 \times 10^{13} \text{ m}$$

The Solar System is roughly 10^{13} meters in diameter, so the AGN emitting region in this example is about 18 times the diameter of the Solar System—an astoundingly small size given that a typical AGN can be a billion times more massive than the Sun and about 1,000 times more luminous than the entire Milky Way Galaxy.

As the light from the back of the region travels through the region, the light may be delayed by interacting with matter in this region, lengthening the variation time. Because of this, the AGN may have faster variations within slower variations. Thus, the fastest variation gives an upper limit to the size of the region: The region cannot be larger than the size calculated from this fastest variation. (It might still be smaller, because you might not have observed the fastest possible variation.)

Imagine that you and your observing team have been monitoring a quasar for about a decade. **Figure 24.3** shows the "data" you obtained between 2000 and 2012 at visual wavelengths. It is difficult to determine the "normal" level of flux for this quasar, as there is always some level of variability. One flare is clearly visible during 2005, with another perhaps starting just before the observations ended.

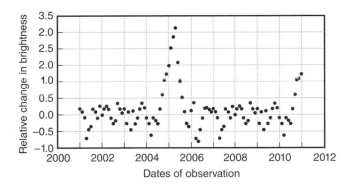

FIGURE 24.3

1. Determine the variation time, Δt, for the 2005 flare by finding the start and end time for the flare.

   Start date: _____ End date: _____ Time elapsed: _____

2. Convert the variation time to seconds and round off your answer. (There are about 3.1×10^7 seconds in a year.)

   $\Delta t =$ _____ seconds

3. Calculate the size of the emitting region, $d = c \times \Delta t$. (Round your answer to two significant digits.)

   Size of emitting region: _____ meters

4. The Solar System is $\sim 10^{13}$ meters in diameter. What is the size of the emitting region in units of Solar Systems?

   _____ Solar Systems

Step 2—Measuring the Energy Emitted per Second by an Active Galaxy Flare

It is possible to determine the total energy emitted during the maximum of a flare and compare it with the energy emitted by the Sun. Remember that the AGN emits light in all directions, so that the total energy emitted by the flare is spread out over a sphere centered on the AGN, with a radius equal to the distance from Earth to the AGN.

5. The distance to this quasar is 2.4×10^9 light-years. This is the radius of the sphere over which the energy is spread out. Convert the units of this radius to meters by multiplying by 9.5×10^{15} meters per light-year (m/ly):

_____ meters

6. The equation for the surface area of a sphere is $A = 4\pi r^2$. Calculate the surface area using the radius of this sphere.

_____ m^2

7. We estimate 1.2×10^{-13} watts per square meter (W/m^2) to be the "normal" (that is, nonflare) power per square meter received at Earth. Multiply this by the area of the sphere at Earth's distance to find the power output of the AGN.

_____ W

8. The maximum flux of the 2005 flare is 3.6×10^{-13} W/m^2. What is the total luminosity of the flare? (Hint: Compare this number with the "normal" level in question 7. How many times greater is the maximum flux from the 2005 flare than the normal level?)

_____ W

9. The Sun's luminosity is about 4×10^{26} W. What is the ratio of the maximum luminosity of the 2005 flare to the Sun's luminosity?

_____

10. Consider the observations and calculations presented here and comment as to whether or not the evidence supports the presence of a supermassive black hole powering your newly discovered quasar. Support your answer.

11. The observations presented in Figure 24.3 are nearly identical to those reported for the quasar 3C 273. Even at 2.4 billion light-years away, this quasar is one of the closest and brightest in the sky. Given the conclusion you reached for your newly discovered quasar, can you infer that a supermassive black hole is powering the quasar 3C 273? Explain.

ACTIVITY 25

RR Lyrae Stars and the Distance to the Center of the Galaxy

Learning Goals

In this activity you will use observations of RR Lyrae variable stars to determine the distances to globular clusters and from that find out where they are located in the Milky Way. You will be determining the size of the Milky Way using a method similar to that done originally. After completing the activity you will be able to:

1. Determine the distances to two globular clusters using RR Lyrae variables found in them.

2. Estimate the distance to the center of the Milky Way based on the distribution of 130 globular clusters.

3. Relate the contributions RR Lyrae variable stars have made to our knowledge of the Milky Way and our location in it.

Step 1—Finding Distances to a Sample of RR Lyrae Variable Stars

To map the globular clusters from three-dimensional space onto a two-dimensional sheet for this activity, we use the projected distances rather than the actual distances. As **Figure 25.1** shows, the actual distances are greater than those given here, but the center of the distribution will still be in the same location.

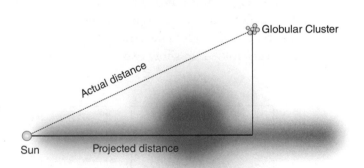

FIGURE 25.1

The variable stars listed in Tables 25.1 and 25.2 are of similar average apparent magnitude; but because they are variable stars, their brightness changes over time. To find the average apparent magnitude for each star, draw a horizontal line across the top of each light curve that measures the brightest magnitude. Draw another horizontal line across the bottom of each light curve that measures the dimmest magnitude.

Add these magnitudes together, and divide by 2 to find the average apparent magnitude. **Figure 25.2** gives an example. The top horizontal line crosses the brightest magnitude: 14.25. The bottom horizontal line crosses the dimmest magnitude: 15.05. The average apparent magnitude is then (15.05 + 14.25)/2, or 14.65.

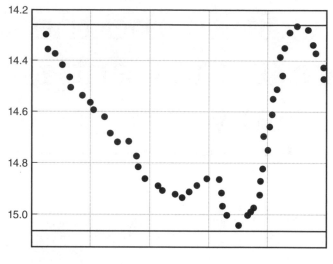

FIGURE 25.2

Figure 25.3 shows light curves of six stars out of a total of 45 observed by astrophysicists Silbermann and Smith in the globular cluster M15. (Although observations were taken through multiple filters, we show only the visual, *V*, filter observations for this activity.) **Figure 25.4** shows light curves of eight stars out of a total of 12 observed by astrophysicists Kauzny, Krzeminski, and Mazur in the globular cluster Ruprecht 106 (RU 106). In both data sets, the *y*-axes are a measure of the apparent magnitude. The *x*-axes show the "phase-folded" data, where the observations over time have been "folded" to show two or more complete cycles. The light curves look similar, but

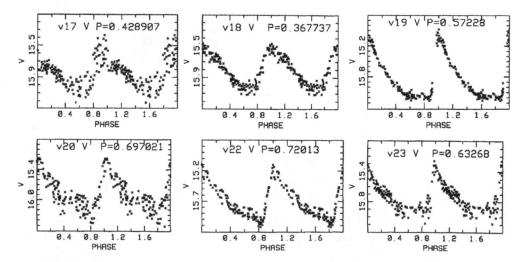

FIGURE 25.3

there are subtle differences depending on the type of RR Lyrae star. All of them show distinctive changes in brightness over time.

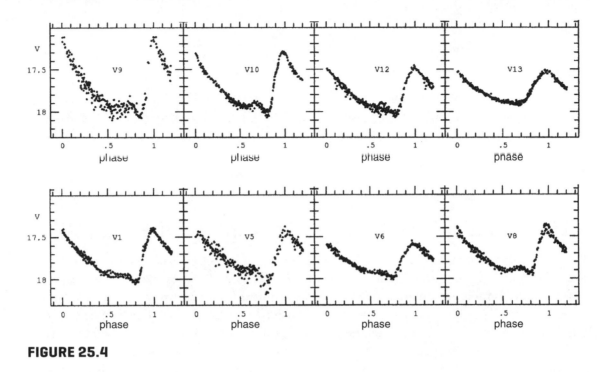

FIGURE 25.4

1. Find the average magnitude of each star in Figures 25.3 and 25.4 and fill in **Tables 25.1 and 25.2**.

⊙ TABLE 25.1

RR Lyrae stars in M15.

VARIABLE STAR ID	BRIGHTEST MAGNITUDE	DIMMEST MAGNITUDE	AVERAGE MAGNITUDE
v17			
v18			
v19			
v20			
v22			
v23			
Average magnitude for M15			

⊙ TABLE 25.2

RR Lyrae stars in RU 106.

VARIABLE STAR ID	BRIGHTEST MAGNITUDE	DIMMEST MAGNITUDE	AVERAGE MAGNITUDE
v9			
v10			
v12			
v13			
v1			
v5			
v6			
v8			
Average magnitude for RU 106			

Because we have average values for both the apparent magnitudes for each set of stars and published average absolute magnitude 0.59 for RR Lyrae stars, we can use a version of the magnitude equation to solve for the distance in parsecs (pc):

$$d = 10^{[(m-M+5)/5]}$$

2. Use the distance equation to find the distance to the nearer cluster: _____ pc.

3. Use the distance equation to find the distance to the farther cluster: _____ pc.

4. The uncertainty in the published value for the absolute magnitude is ± 0.03. How would the distances you calculated to the two globular clusters change if you used the largest value allowed by the uncertainty ($M = 0.62$)?

Step 2—Finding the Distance to the Center of the Milky Way

Figure 25.5 is a plot of 130 of the ~150 known globular clusters that exist as part of the Milky Way. (Galactic longitude and distances are from W. E. Harris, 1996, *Astronomical Journal*, vol. 112, p. 1487; there is more information on these clusters at http://physwww.physics.mcmaster.ca/~harris/mwgc.dat.) The globular clusters are indicated by the black dots. The distances from the Sun have been rounded off to the nearest kiloparsec (kpc; 1 kpc = 1,000 pc, or 3,260 light-years), making multiple clusters appear to be at the same distance. The dots at the edges of the graph represent clusters that are farther away than 19 kpc. The farthest cluster is at a projected distance of 36 kpc.

Examine Figure 25.5. The Sun and thus Earth are at the very center of this graph.

5. Estimate the center of the distribution of globular clusters. The best way to do this is to use a small circular object about the size of a quarter or water bottle cap and try to include as many globular clusters as possible within the circumference of the object you use. Make an "X" on the graph at the center of the distribution.

6. Estimate the distance from the Sun to the center of the distribution.

Distance = _____ kpc

7. Estimate the radius of the galaxy based on the overall full extent of the distribution of the clusters.

Radius = _____ kpc

8. Determine the direction to the center of the distribution. This is considered to be the direction to the center of the galaxy.

Longitude to center of the distribution = _____ degrees

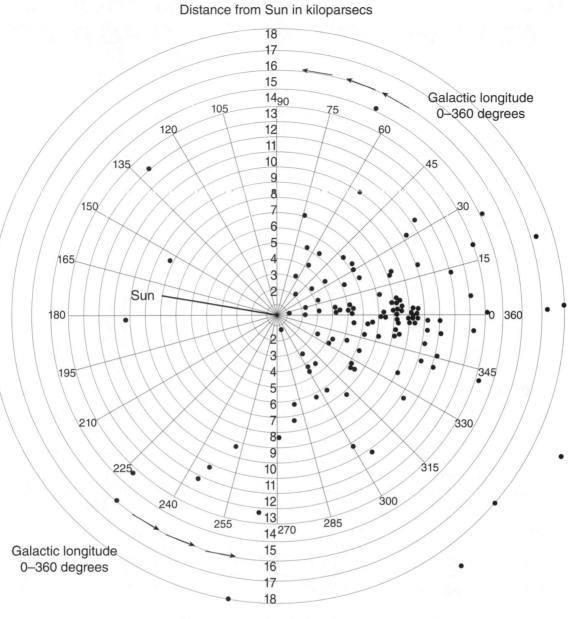

FIGURE 25.5

9. Based on this information, explain how we know the Sun is not at the center of the Milky Way Galaxy.

10. The galactic longitude of M15 is 65°; the galactic longitude of RU 106 is 301°. Based on these coordinates and the distances you found above, mark the location of both globular clusters on Figure 25.5.

11. Mark the location of the globular clusters in **Table 25.3** on Figure 25.5. Do these added data points change any of your previous conclusions about the center of the galaxy or our distance from the center? Explain.

12. Based on what you have learned through this activity, relate the contributions RR Lyrae variable stars have made to our knowledge of the Galaxy and our location in it.

● TABLE 25.3

The galactic longitude and distance to nine globular clusters in the Milky Way.

CLUSTER ID	GALACTIC LONGITUDE (DEGREES)	DISTANCE (KPC)	CLUSTER ID	GALACTIC LONGITUDE (DEGREES)	DISTANCE (KPC)	CLUSTER ID	GALACTIC LONGITUDE (DEGREES)	DISTANCE (KPC)
NGC 6093	353	10	NGC 6101	318	15	NGC 6121	351	2
NGC 6139	342	10	NGC 6144	352	9	NGC 6453	356	10
NGC 6496	348	12	NGC 7089	53	12	NGC 7492	53	26

⬤ ACTIVITY 26

Finding the Expansion Rate and the Age of the Universe

Learning Goals

In this activity, you will analyze the distances to galaxies and the velocities at which they are receding from us. By the end of this activity you will be able to:

1. Determine a value for the Hubble constant through analysis of measurements of the angular sizes and redshifts of galaxies.

2. Estimate the corresponding age of the universe and compare it with the age of the Sun and the Milky Way.

3. Explain why Hubble's law implies that the universe is expanding.

4. Summarize how our view of the universe has changed as the value of the Hubble constant has improved.

Step 1—An Analysis of a Sample of Galaxies

The graph of the velocity of each galaxy in kilometers per second versus its distance in megaparsecs gives us enough information to find the best-fit slope and thus Hubble's constant. Once you have found Hubble's constant, it is just a matter of taking its inverse and adjusting the units to find the age of the universe. We will start with a closer look at a sample of galaxies. Look for similarities among the galaxies and among the features of their spectra in **Table 26.1**.

⬤ TABLE 26.1

Images, spectra, identifications, redshifts, recessional velocities, and distances to 18 galaxies.

GALAXY IMAGE	GALAXY SPECTRUM	GALAXY ID	REDSHIFT	VELOCITY (km/s)	DISTANCE (Mpc)
		NGC 1357	0.006651	1,994	25

(continued)

○ TABLE 26.1

Continued.

GALAXY IMAGE	GALAXY SPECTRUM	GALAXY ID	REDSHIFT	VELOCITY (km/s)	DISTANCE (Mpc)
	NGC 1832	NGC 1832	0.00646	1,937	32
	NGC 2276	NGC 2276	0.00806	2,417	26
	NGC 2775	NGC 2775	0.00451	1,353	20
	NGC 2903	NGC 2903	0.00186	556	10
	NGC 3034	NGC 3034	0.00073	219	12
	NGC 3147	NGC 3147	0.00935	2,804	22

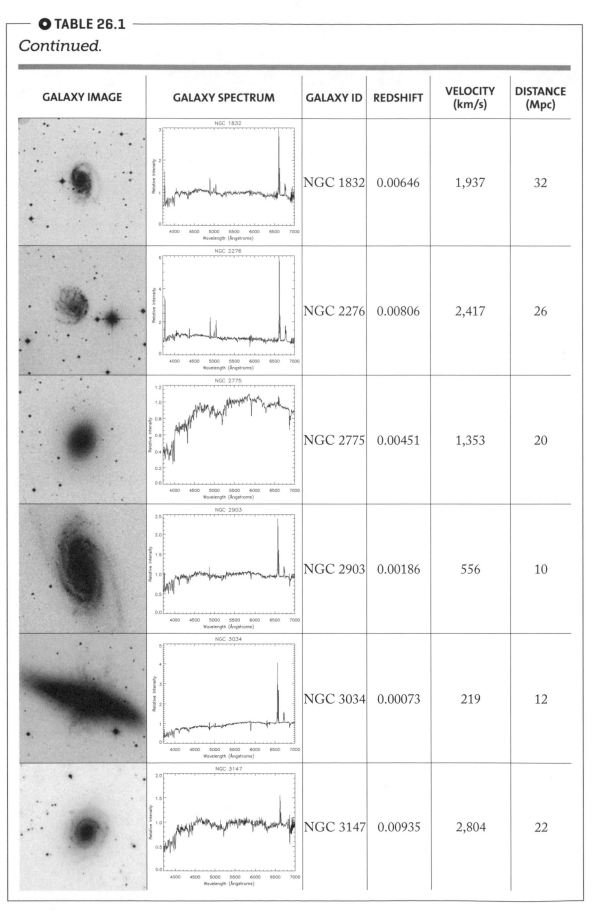

⊙ TABLE 26.1

Continued.

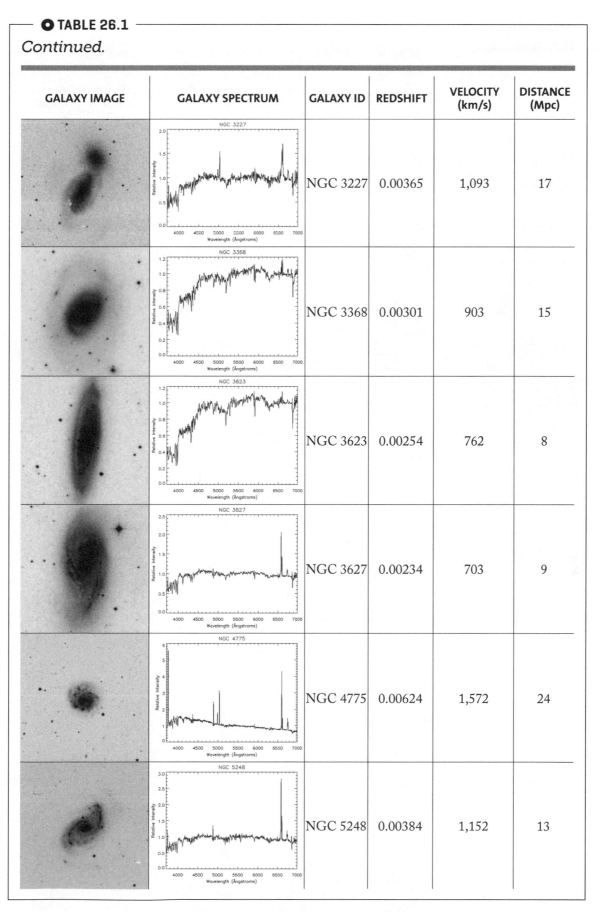

GALAXY IMAGE	GALAXY SPECTRUM	GALAXY ID	REDSHIFT	VELOCITY (km/s)	DISTANCE (Mpc)
	NGC 3227	NGC 3227	0.00365	1,093	17
	NGC 3368	NGC 3368	0.00301	903	15
	NGC 3623	NGC 3623	0.00254	762	8
	NGC 3627	NGC 3627	0.00234	703	9
	NGC 4775	NGC 4775	0.00624	1,572	24
	NGC 5248	NGC 5248	0.00384	1,152	13

(continued)

● TABLE 26.1

Continued.

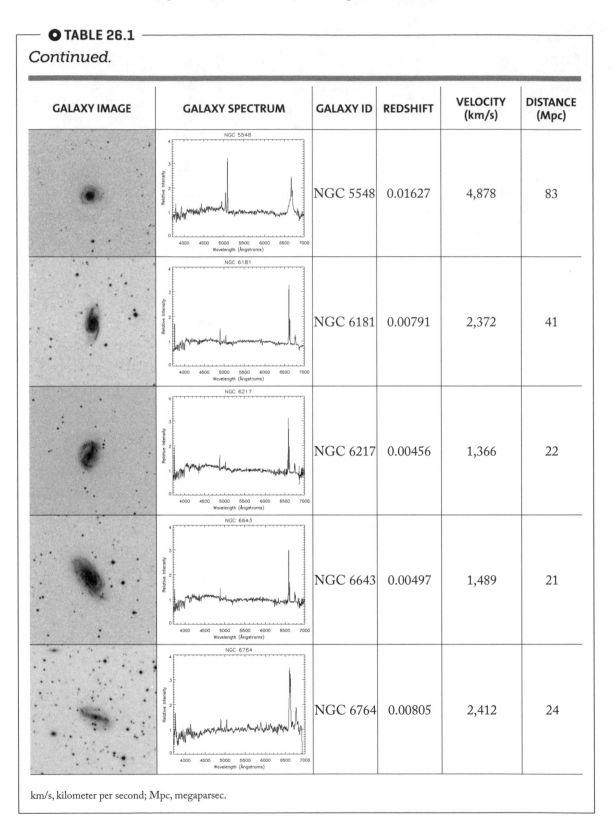

GALAXY IMAGE	GALAXY SPECTRUM	GALAXY ID	REDSHIFT	VELOCITY (km/s)	DISTANCE (Mpc)
	NGC 5548	NGC 5548	0.01627	4,878	83
	NGC 6181	NGC 6181	0.00791	2,372	41
	NGC 6217	NGC 6217	0.00456	1,366	22
	NGC 6643	NGC 6643	0.00497	1,489	21
	NGC 6764	NGC 6764	0.00805	2,412	24

km/s, kilometer per second; Mpc, megaparsec.

1. Recall Hubble's tuning fork for the classification of galaxies. What is the general classification for the galaxies of Table 26.1?

2. Astronomers often use the concept of a standard ruler—objects that look similar most likely are approximately the same actual size—for determining relative distances to objects. Why is it important to have all of the galaxies we are examining here be of the same or very similar type?

3. We can add additional criteria to increase confidence that these galaxies can be used as standard rulers. Examine the spectra for the galaxies. There are strong features—emission and absorption—that can be seen at similar wavelengths in each spectrum. Describe some of these emission and absorption lines.

4. The images of the galaxies shown in Table 26.1 were taken by the same large, ground-based telescope using the same instruments and equipment, and so are all at the same scale. Each image measures 0.12 degrees on each side. **Figure 26.1** shows enlarged images of two galaxies that look very similar: (a) is NGC 3368 and (b) is NGC 3147. List two possible reasons why they are not the same angular size. Can you determine which of those reasons is the right one? Explain.

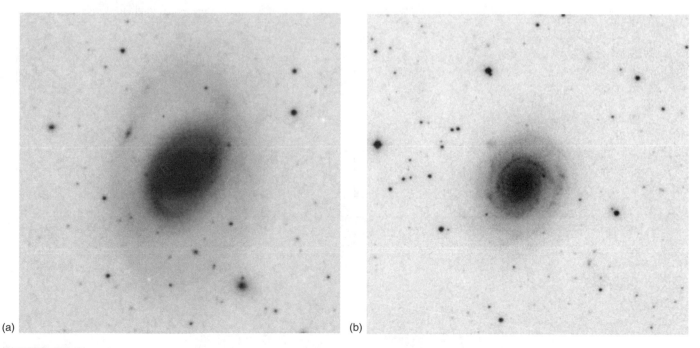

(a)   (b)

FIGURE 26.1

Step 2—Finding the Hubble Constant and Calculating the Age of the Universe

5. **Figure 26.2** shows the spectra for the two galaxies of Figure 26.1. Examine the spectra for these two galaxies. The spectra for NGC 3147 are (a) and (b). The spectra for NGC 3368 are (c) and (d). These spectra highlight the regions around the absorption lines (historically called calcium H and K lines) for the singly ionized calcium atoms in the stars in these galaxies and for the Balmer line H-alpha seen in emission, perhaps coming from star-forming regions. Which galaxy is farther away? How do you know?

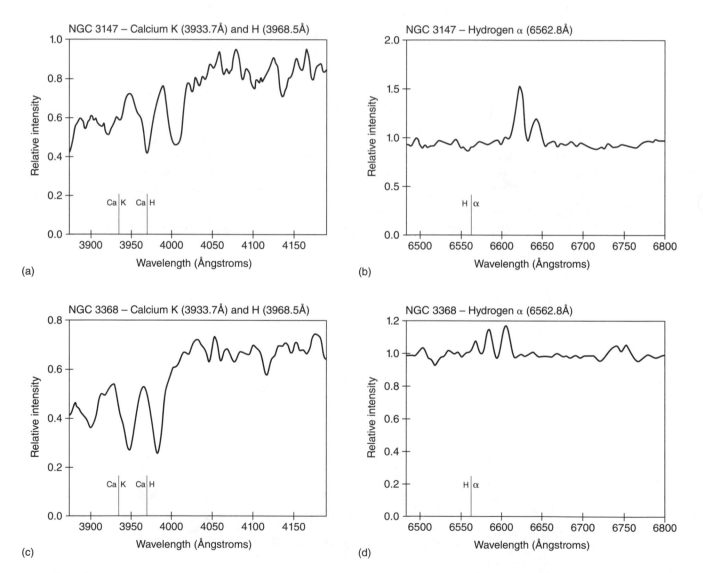

(a)

(b)

(c)

(d)

FIGURE 26.2

6. Check the redshift values for these two galaxies, as given in Table 26.1. Approximately how much farther away is the more distant galaxy from the closer one?

7. Now compare the distances to these two galaxies as determined by the standard ruler method. Did you discover an inconsistency between the relative distances obtained from the redshifts of NGC 3147 and NGC 3368 and the relative distances calculated by the standard ruler method? Which comparison is likely to be more accurate? Why?

Figure 26.3 shows the graph of the measured recessional velocities versus the measured distances to the 18 galaxies in this sample. The line b is the approximate best fit to the data; lines a and c were obtained by ignoring a couple of outliers in the measurements. The equation of the line is in the form $y = mx + b$, where y equals the velocity, x is the distance, and the slope, m, will be the value of the Hubble constant. The y-intercept, b, equals zero.

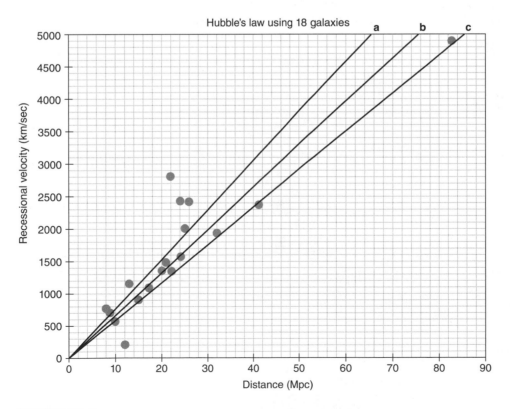

FIGURE 26.3

8. Why must the *y*-intercept be equal to zero? (Hint: What is at zero distance from us, and how fast is it moving?)

9. Calculate the slope of the best-fit line, b, and round off the number to two significant digits (for example, 75.4839485 would be 75). Then calculate the slopes of the steepest allowed fit, a, and the flattest allowed fit, c.

Slopes: Best-fit line b: _____ km/s/Mpc

Steepest allowed a: _____ km/s/Mpc

Flattest allowed c: _____ km/s/Mpc

10. Which line will give the youngest age for the universe? _____

11. Which line will give the oldest age for the universe? _____

We can now calculate the age of the universe based on these data. If the universe has been expanding at a constant speed since its beginning, its age would simply be the inverse of the Hubble constant, $1/H_0$. However, the units of the Hubble constant, kilometers per second per megaparsec (km/s/Mpc), need to be converted so that the inverse has units of time.

Example: Let's say you find a Hubble constant of 75 km/s/Mpc, then:

$$\frac{1}{75 \text{ km/s/Mpc}} = 0.0133 \frac{\text{s} \cdot \text{Mpc}}{\text{km}}$$

$$0.0133 \frac{\text{s} \cdot \text{Mpc}}{\text{km}} = 1.33 \times 10^{-2} \text{ s} \frac{\cdot \text{Mpc}}{\text{km}}$$

$$1.33 \times 10^{-2} \frac{\text{s} \cdot \text{Mpc}}{\text{km}} \times 3.09 \times 10^{19} \frac{\text{km}}{\text{Mpc}} = 4.12 \times 10^{17} \text{ s}$$

$$4.12 \times 10^{17} \text{ s} \times \frac{1 \text{ year}}{3.16 \times 10^7 \text{ s}} = 1.3 \times 10^{10} \text{ years}$$

This is 1.3×10^{10} years, or 13×10^9 years, which is 13 billion years.

12. Now, carry out the analysis for your value of H_0. First, find the inverse of your value of H_0.

Inverse: _____

13. Second, multiply the inverse by 3.09×10^{19} km/Mpc to cancel the distance units.

Age: _____ sec

14. As you now have the age of the universe in seconds, divide this number by the number of seconds in a year: 3.16×10^7 s/yr.

Age: _____ years

15. Quantitatively (use ratios) compare your maximum age for the universe to the age of the Sun (5 billion years) and to the age of the oldest stars in the Milky Way (approximately 12.5 billion years). Comment on your findings.

16. The long-standing view of the universe before Edwin Hubble's observations was that everything was standing still. Discuss how your analysis either supports or refutes this claim.

When Hubble made his first observations, he determined a Hubble constant of 500 km/s/Mpc. During the years 1950 to 1960, there were serious disagreements about the value for the Hubble constant. One group stated that it was 60 km/s/Mpc; another group said it was closer to 100 km/s/Mpc. The current results from the Planck mission give the best estimate of the Hubble constant: 67.8 ± 0.77 km/s/Mpc (see http://map.gsfc.nasa.gov/universe/bb_tests_exp.html).

17. Summarize how our knowledge of the universe has changed with the more exact value of today.

ACTIVITY 27

A Cosmic Calendar

Learning Goals

In this activity you will organize the natural history of the universe onto a calendar representing one Earth year in order to gain a feel for the relative magnitudes of the timescales involved. During the activity you will need to:

1. Calculate relative timescales.

2. Identify important events in natural history.

3. Choose appropriate time steps at different epochs.

Step 1—An Overview of Some of the Events in the History of the Universe

● TABLE 27.1

Some events in the history of the universe.

	COLUMN 1	COLUMN 2	COLUMN 3	COLUMN 4	COLUMN 5	COLUMN 6
	EVENT	ACTUAL TIME (YEARS AGO)	1 – (FRACTION OF YEAR)	DAY NO.	DATE	TIME
Big Bang		13.8×10^9	0	1	January 1	—
Milky Way formed		11×10^9	0.2	73	March 14	—
Sun formed		4.6×10^9		245		—
Oldest known Earth rocks		4.0×10^9	0.71	259	September 16	—
First (known) life		3.8×10^9	0.72	264	September 21	—
Photosynthesis		3×10^9	0.78		October 13	—
Oxygenation of atmosphere		2.4×10^9	0.83	302	October 29	—
Eukaryotes		2×10^9	0.86	312	November 8	—
Multicellular life		1×10^9	0.93	339		—
Simple animals		0.67×10^9	0.95	347	December 13	—
Fish		0.5×10^9	0.96	352	December 18	—
Amphibians		0.36×10^9	0.97	355	December 20	—
Mammals		0.2×10^9	0.986	360	December 25	—
Birds		0.15×10^9	0.989	361	December 27	—
Flowers		0.1×10^9		362	December 28	—

⊙ **TABLE 27.1**

Continued.

COLUMN 1	COLUMN 2	COLUMN 3	COLUMN 4	COLUMN 5	COLUMN 6
EVENT	ACTUAL TIME (YEARS AGO)	1 − (FRACTION OF YEAR)	DAY NO.	DATE	TIME
Primates	65×10^6	0.995	363	December 29	—
Apes	15×10^6	0.9989	364.60	December 31	14 h 24 m
Hominids	12.3×10^6	0.9991	364.67	December 31	16 h 4.8 m
Primitive humans	2.5×10^6	0.9998	364.93	December 31	
Domestication of fire	0.4×10^6	0.99997		December 31	23 h 44 m
Most recent ice age begins	0.11×10^6	0.999992	364.997	December 31	23 h 56 m
Sculpture and painting	0.035×10^6	0.999997	364.999	December 31	23 h 59 m
Agriculture	0.012×10^6	0.99999913	364.99968	December 31	23 h 59 m 32 s
First writing	5.5×10^3	0.99999960	364.99985	December 31	23 h 59 m 47 s
The wheel	4.5×10^3	0.99999967	364.99988	December 31	23 h 59 m 50 s
Roman Republic	2.5×10^3	0.99999982	364.999934	December 31	23 h 59 m 54 s
Renaissance	1.0×10^3	0.999999928	364.999974	December 31	
Modern science	4×10^2	0.999999971	364.999989	December 31	23 h 59 m 59.09 s
Your birth	$\sim 2 \times 10^1$	0.999999999	364.9999995	December 31	

h, hour; m, minute; s, second.

⊙ **TABLE 27.2**

Day number for the first day of each month.

DATE	DAY NO.	DATE	DAY NO.	DATE	DAY NO.	DATE	DAY NO.
January 1	1	April 1	91	July 1	182	October 1	274
February 1	32	May 1	121	August 1	213	November 1	305
March 1	60	June 1	152	September 1	244	December 1	335

Step 2—Scaling Dates and Times

1. You are familiar with scaling relations from maps, which transform one length scale into another in order to "de-magnify" reality so that it fits onto a piece of paper. This makes finding the relationships between different locations much simpler. We can perform a similar trick with universal time scales, scaling the entire age of the universe onto a Cosmic Calendar year, so that we can more easily understand the relationships between the time intervals. To understand this, you must first fix in your head the meaning of "a year."

   a. A lot happens in a year. Take a moment to think about the past year. List two or three significant events that have happened to you in the past year.

b. Now take a moment to think about the past month. List two or three events from the past month that are memorable.

c. Think about the past day. List two or three things that happened in the past day that caught your attention.

d. For the past minute, you have been focused on this assignment. But take a moment to think about something significant that can happen between one minute and the next, and write it down, even if it didn't happen to you THIS minute.

e. Think of something significant that can happen between one second and the next, and write it down, even if it didn't happen to you THIS second.

2. Study **Table 27.1**, which lists a number of events in the history of the universe, and the scaling relationship onto a year for each of these time points. Column 3 gives the fraction of the year since the beginning at which the event happens. These numbers come from dividing the time of the event in units of "years ago" by the total age of the universe, 13.8×10^9 years and subtracting that number from 1. Fill in the empty spots in column 3 of Table 27.1.

3. Column 4 gives the day number that corresponds to the scaled time of the event. This is calculated by multiplying the number in column 3 by 365. Fill in the empty spots in column 4 of Table 27.1.

4. Column 5 shows the date corresponding to the day number from column 4. Because each month contains a different number of days, this is a little tricky. **Table 27.2** contains the day number of the first day of each month, to help you with the conversion from day number to date. Fill in the empty spots in column 5 of Table 27.1. [Hint: Besides using Table 27.2, you may also choose to count days ahead or days behind of the month and day that brackets each empty spot.]

Step 3—Filling in the Last Day of this Cosmic Calendar

Column 6 shows the time (in hours, minutes, and seconds) that corresponds to the fractional day in column 4. This is only necessary on the last day of the year because so many events happened so close together in time. Time is not measured in base 10, and instead has 60 seconds in a minute, 60 minutes in an hour, 24 hours in a day, and 365 days in a year. This makes the conversion from fractional days to hours, minutes, and seconds cumbersome. To make this conversion, one must make some calculations as the following examples show.

    a. Find the fractional day by subtracting 364 from the number in column 4. For example, agriculture began on day number 364.99968. This is near the end of the day on December 31 (the day after the 364th day). How far through this day? The rise of agriculture occured 0.99968 of the way through the day.

    b. Multiply this fraction of a day by 24 hours to find out how many hours are accounted for in this fraction. In the case of the rise of agriculture, $0.99968 \times 24 = 23.99232$ hours. Agriculture arose in the last hour of the last day.

    c. To find the number of minutes of the last hour of the last day, we must take the fractional hour (0.99232) and multiply it by 60, which gives: 59.5392 minutes. Agriculture arose on December 31, in the 23rd hour 59th minute.

    d. To find the number of seconds in the last minute of the last hour of the last day, we must take the fractional minute (0.5392) and multiply it by 60, which gives: 32.352. Depending on the times around this time, we might retain the fractional second or we might not. In this case, the events on either side of the rise of agriculture are many seconds earlier or later, so we can ignore the fractional second. The rise of agriculture occurred on December 31 at 23 h 59 m 32 s.

5. Choose two events from Table 27.1, and answer the following questions.

    a. What are the two events?

    b. How long ago (on the Cosmic Calendar 1-year timescale) did each event occur?

    c. How much time elapsed between them?

    d. What fraction of the Cosmic Calendar year is the time from part (c)?

6. Fill in the empty spots in column 6 of Table 27.1. Show an example of one of your calculations here.

To make these numbers more clearly understandable, it's helpful to place the events onto our more usual time-measuring devices. Place all of the events in Table 27.1 onto **Figures 27.1 to 27.4**, switching from one figure to the next when necessary to space out the information you are adding to the diagrams.

January

			1	2	3	4
5	6	7	8	9	10	11
12	13	14	15	16	17	18
19	20	21	22	23	24	25
26	27	28	29	30	31	

February

						1
2	3	4	5	6	7	8
9	10	11	12	13	14	15
16	17	18	19	20	21	22
23	24	25	26	27	28	

March

						1
2	3	4	5	6	7	8
9	10	11	12	13	14	15
16	17	18	19	20	21	22
23	24	25	26	27	28	29
30	31					

April

	1	2	3	4	5
6	7	8	9	10	11	12
13	14	15	16	17	18	19
20	21	22	23	24	25	26
27	28	29	30			

May

			1	2	3
4	5	6	7	8	9	10
11	12	13	14	15	16	17
18	19	20	21	22	23	24
25	26	27	28	29	30	31

June

1	2	3	4	5	6	7
8	9	10	11	12	13	14
15	16	17	18	19	20	21
22	23	24	25	26	27	28
29	30					

July

	1	2	3	4	5
6	7	8	9	10	11	12
13	14	15	16	17	18	19
20	21	22	23	24	25	26
27	28	29	30	31		

August

					1	2
3	4	5	6	7	8	9
10	11	12	13	14	15	16
17	18	19	20	21	22	23
24	25	26	27	28	29	30
31						

September

		1	2	3	4	5	6
7	8	9	10	11	12	13
14	15	16	17	18	19	20
21	22	23	24	25	26	27
28	29	30				

October

			1	2	3	4
5	6	7	8	9	10	11
12	13	14	15	16	17	18
19	20	21	22	23	24	25
26	27	28	29	30	31	

November

						1
2	3	4	5	6	7	8
9	10	11	12	13	14	15
16	17	18	19	20	21	22
23	24	25	26	27	28	29
30						

December

See Figure 27.2

FIGURE 27.1

December						
	1	2	3	4	5	6
7	8	9	10	11	12	13
14	15	16	17	18	19	20
21	22	23	24	25	26	27
28	29	30	31			

FIGURE 27.2

FIGURE 27.3

The last minute of the year: December 31, 23 hr 59 m

FIGURE 27.4

7. Astronomers (and other scientists) often rescale sizes, distances, or times as you have done in this activity. Why is this a useful skill to have?

ACTIVITY 28

The Hubble Deep Field North

Learning Goals

In this activity, you will use the Hubble Deep Field North to explore the evolution and distribution of galaxies in the universe. As you work through the activity you will:

1. Analyze the galaxies in and the structure of the Hubble Deep Field North.

2. Identify Hubble types and relate types to colors of galaxies.

3. Relate the colors of galaxies to the possible dominant stellar populations.

4. Summarize the importance of the Hubble Deep Field North to our understanding of the evolution of galaxies.

Step 1—Galaxy Types

In this activity, you will examine the Hubble Deep Field North (HDFN), an image of very distant space as it was far back in time. After this image was obtained, the 10-meter, Earth-based Keck telescope was used to observe the faint blue galaxies in the image. Astronomers have concluded that the small blue shards are among the most distant objects ever seen. See if you can pick out some of these faint blue galaxies in the image that your instructor is projecting. These objects may represent galaxies caught in the act of formation. In all, the number of galaxies in the image implies that there are about 40 billion galaxies in the observable universe.

Take a close look at **Figure 28.1**, the HDFN. Noted next to many of the galaxies is the redshift, also known as z, for that galaxy. You will need those redshifts in Step 2 of this activity. There are a number of clearly identifiable galaxy types included in the HDFN. You may want to refer to the source image as you work: http://ned.ipac.caltech.edu/level5/Deep_Fields/mirror/hdfn/index.html.

1. Make your own map of the Hubble Deep Field North by locating 15–20 galaxies. Sketch them, placing them on the grid of **Figure 28.2** according to their locations in the image.

2. Label each galaxy sketched with the Hubble type of that galaxy, and create a key on the grid for identification. (Clicking on a galaxy in the online source image gives a "close-up" view through five different filters that will help in the classifying.)

3. The galaxies have noticeably different colors, as you have noted in the image source. Do you see any overall pattern between the color of the galaxy and its Hubble type? Based on what you know about the colors of stars, give a brief description of the types of stars that make up the different types of galaxies.

FIGURE 28.1

Step 2—Redshifts

Return to Figure 28.1. You will now examine the redshifts of the galaxies in the image. The corresponding galaxy is the galaxy located closest to the redshift number. The redshift is defined as the fraction of the speed of light that the galaxy is receding from us, v/c, as measured by the observed change in wavelength of lines in its spectrum:

$$\frac{\lambda_{observed} - \lambda_{rest}}{\lambda_{rest}} = \frac{v}{c} = z$$

Start by assuming the HDFN represents the actual distribution of galaxies in this portion of the sky. To determine if there is structure in the image, you need to use the histogram of the cosmic redshifts, **Figure 28.3**. This histogram contains data for almost all of the galaxies in this field, even those that look like little specks. For this histogram, the redshifts have been binned from $z = 0.0$ to $z = 4.2$ in increments of 0.2. The curvy line represents an average of the previous two data points and emphasizes that there seems to be "groupings" of galaxies around certain redshifts.

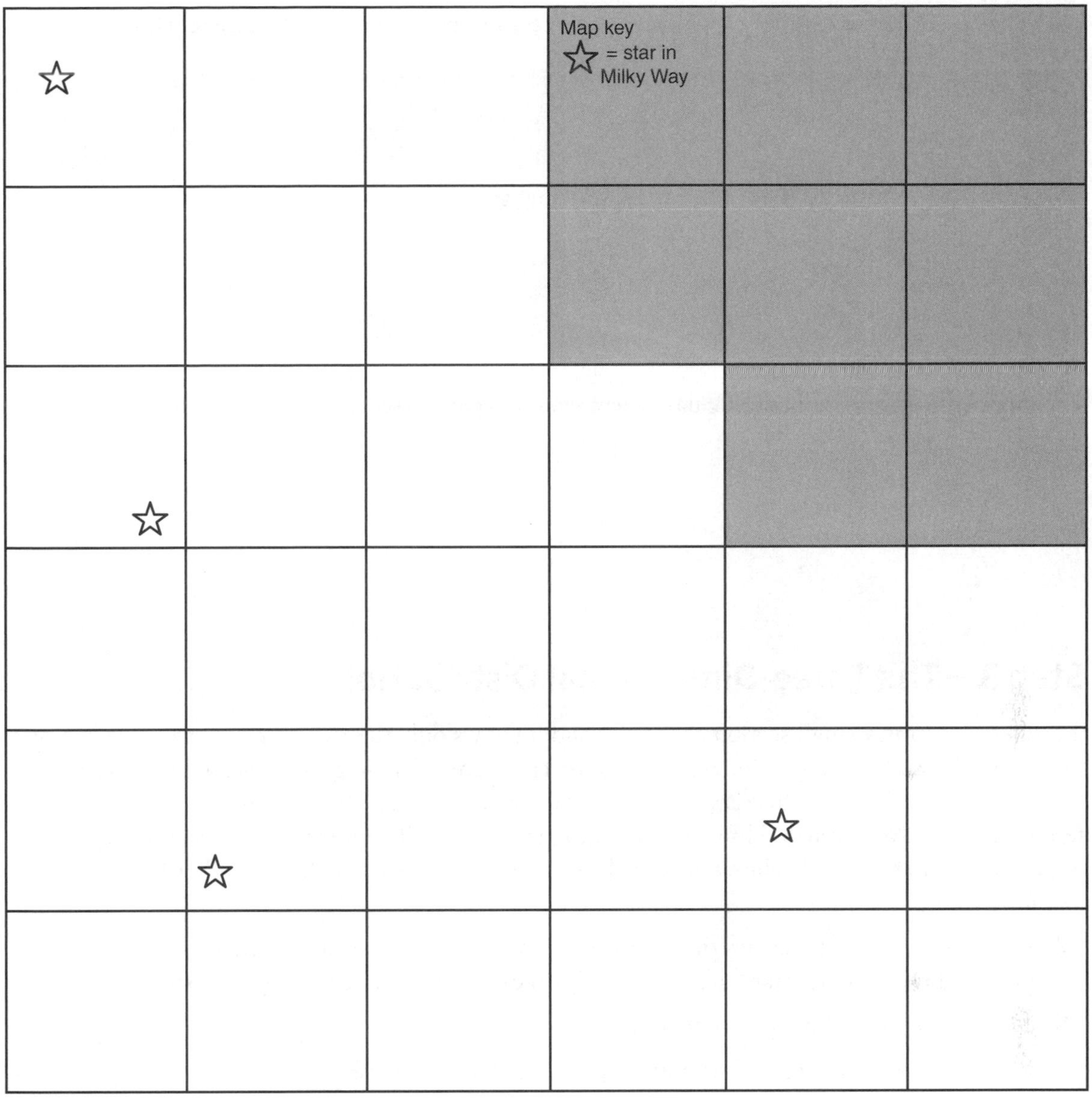

Map key
☆ = star in
Milky Way

FIGURE 28.2

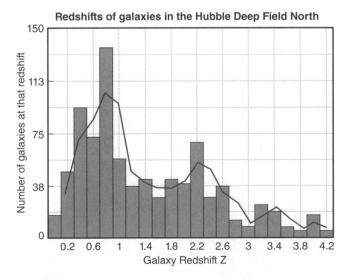

Redshifts of galaxies in the Hubble Deep Field North

Number of galaxies at that redshift

150

113

75

38

0

0.2 0.6 1 1.4 1.8 2.2 2.6 3 3.4 3.8 4.2

Galaxy Redshift Z

FIGURE 28.3

4. There are three redshifts for which galaxies seem to be grouped. List those redshifts.

5. At what redshifts do there seem to be very few galaxies?

6. Suppose that you had a bigger, more powerful telescope, and looked even deeper into space. Speculate on the number of galaxies you would expect to see at redshifts greater than 4.2.

Step 3—The Three-Dimensional Distribution

This is an extremely small section of the sky, and it shows only the two-dimensional distribution of galaxies. From this image alone, it is not apparent if any of these galaxies are actually near one another in space. Galaxies that appear close together on the image might actually be at very different distances from Earth. However, by adding in the redshift information, we can find groups of galaxies that are actually close together. These groups will be both close together in the image *and* have similar redshifts.

7. Look for a group of galaxies that you would classify as a cluster, and sketch four or more galaxies belonging to that "cluster" on the grid of Figure 28.2. Circle your cluster.

8. Defend your decision, quantitatively.

    a. By how much does the redshift vary among these galaxies?

    b. Suppose you noted clustering at redshifts of 0.20 and also at 0.40. By how much does the velocity differ between these clusters of galaxies? Use 300,000 km/s as the speed of light.

    c. Use a Hubble constant of 75 km/s/Mpc and Hubble's law ($v = H_0 \times d$) to calculate the distances to the clusters at redshifts of 0.20 and 0.40.

Step 4—Galaxy Evolution

9. Astronomers examine the images of the galaxies in the Hubble Deep Field North because there are galaxies that show the universe at a relatively early age. Astronomers can compare these galaxies, which are newly formed, to nearer galaxies, which are much older, to understand the changes in galaxies over time. Use the concept of look-back time and explain why studying the Hubble Deep Field North should give us insight as to the evolution of galaxies.

Name _____ Date _____ Section_____

Calculating the Mass of the Central Object of the Galaxy

Learning Goals

In this activity you will use observations of stars orbiting near the center of the Milky Way to confirm the existence of a supermassive black hole as the central object. The method you will use is closely connected to the methods used with orbiting objects in previous activities. After completing the activity you will be able to:

1. Demonstrate proficiency in the use of scientific notation in all calculations.

2. Apply Newton's form of Kepler's third law to infer the mass within the orbit of a centrally located star in the galaxy.

3. Compare the results to the mass of the Sun and the size of the Solar System.

4. Analyze the evidence for a supermassive black hole at the center of the Milky Way.

5. Compare results to those published in astronomical journals.

Step 1—Finding the Mass and Semimajor Axis of the Central Object

Your instructor will show a short video made from infrared observations of stars orbiting the center of the Milky Way. The observations spanned an 18-year time period.

1. Finding the period, P: During the 18 years that the galactic center was observed, the star S0-2 completed a full orbit having a period of 16.17 years. Convert this period to seconds (1 year = 3.16×10^7 seconds).

                                        Orbital period = _____seconds

2. Finding the semimajor axis, a: The scale of the images used in the video about the stars orbiting close to the galaxy's center is shown as 0.05 arcseconds (arcsec). Using this information and measurements of the length of the major axis of the orbit of star S0-2 in millimeters (mm), we find that the angular size of the orbit of this star is 0.165 arcsec. Here are the additional steps needed to find a:

   a. Convert the angular size of the orbit in arcseconds to radians (rad).

   $$\frac{0.165 \text{ arcsec}}{206{,}265 \text{ arcsec/rad}} = \underline{\hspace{1cm}} \text{ rad}$$

   b. The distance to the center of the galaxy is given as 7,940 parsecs (pc). Convert this distance to meters.

   $$7{,}940 \text{ pc} \times 3.09 \times 10^{16} \text{ m/pc} = \underline{\hspace{1cm}}\text{meters}$$

   c. Find the major axis of the orbit using the small-angle formula: Size = angular size × distance.

Size of the orbit = _____ meters

   d. Find the size of the semimajor axis by dividing the major axis by 2.

a = _____ meters

   e. Finally, recalling the gravitational constant, $G = 6.67 \times 10^{-11}$ m³/kg s² (meters cubed per kilogram seconds squared), find the mass of the object at the center of the galaxy using Newton's form of Kepler's third law:

$$M = \frac{4\pi^2}{G} \cdot \frac{a^3}{P^2}$$

Mass of object at center of galaxy = _____ kg

3. The mass of the Sun is approximately 1.99×10^{30} kg. Express the mass of the object at the center of the galaxy in solar masses.

Mass of object at center of galaxy = _____ times the Sun's mass

4. Compare the size of the semimajor axis of the orbit of star S0-2 to the size of the Solar System, using the distance to the outer edge of the Kuiper Belt of about 55 astronomical units (AU). There are 1.496×10^{11} meters per astronomical unit. Comment on your results.

Step 2—Considering Uncertainties and Comparing Results

Our measurement for the star's semimajor axis assumed that it is not tilted from our point of view. It is entirely possible that the orbital plane of the star is tilted toward or away from us. **Figure 29.1** demonstrates this effect for the possible inclinations of the orbit of star S0-2.

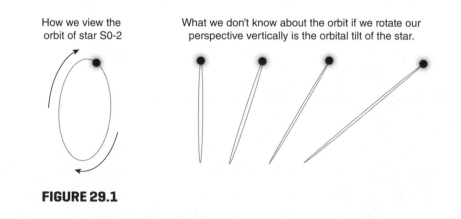

How we view the orbit of star S0-2

What we don't know about the orbit if we rotate our perspective vertically is the orbital tilt of the star.

FIGURE 29.1

5. If the orbit is tilted, will that make the actual semimajor axis larger or smaller than what you measured? Will that make the mass of the central object larger or smaller than your estimate?

6. A number of recent publications give the mass of the central object of the galaxy as approximately 4 million times the mass of the Sun. How do your results compare? Be sure to include comments about the uncertainty of the tilt of the orbit.

7. Astronomers are fairly sure that there is a supermassive black hole at the center of our galaxy. Do you agree or disagree? Support your answer with the results you discovered here.

Name _____ Date _____ Section_____

Habitable Worlds

Learning Goals

In this activity you will explore the properties of known extrasolar planets to determine whether they might support life. During this activity you will also:

1. Describe physical and orbital characteristics of a number of extrasolar planets.

2. Quantitatively compare characteristics of extrasolar planets to the planets in the Solar System.

3. Apply the definition of *habitable zone* to determine habitability of extrasolar planets.

4. Summarize what life might be like on either an exoplanet or a moon of an exoplanet.

Step 1—Stellar Properties

Table 30.1 lists information for just a small sample of the extrasolar planets whose orbits lay all or nearly all in the habitable zone for their respective stars. Information for five Solar System planets is included for comparison.

● TABLE 30.1

Comparison of orbital elements and stellar properties for selected extrasolar planets and five planets of the Solar System.

PLANET NAME (DESIGNATION)	MINIMUM MASS ($M_{JUPITER}$)	RADIUS ($R_{JUPITER}$)	SEMIMAJOR AXIS (AU)	ORBITAL PERIOD (DAYS)	ORBITAL ECCENTRICITY	MASS OF STAR (M_{SUN})	TEMPERATURE OF STAR (K)
Mercury	0.0002	0.03	0.4	88	0.21	1	5778
Earth	0.003	0.09	1.0	365	0.0167	1	5778
Mars	0.0003	0.05	1.5	687	0.09	1	5778
Jupiter	1	1	5.2	4,270	0.05	1	5778
Saturn	0.3	0.8	9.6	10,759	0.06	1	5778
HD 10180 g	0.067	—	1.42	602	0.00	1.06	5911
HD 99109 b	0.50	—	1.11	439	0.09	0.94	5272
HD 28185 b	5.8	~1*	1.02	379	0.05	0.99	5656
HD 73534 b	1.07	~1	3.02	1,770	0.07	1.17	4884
HD 183263 b	3.57	~1	1.49	626	0.36	1.12	5936
55 Cnc f	0.173	—	0.77	261	0.32	0.9	5196

*It is hypothesized that Jupiter's radius is about as large as a gaseous planet can get because additional mass will tend to squeeze the planet more. For the planets in this sample with masses lower than that of Jupiter, no radius is known because they do not transit their stars.

---

1. Quantitatively compare the range in surface temperatures of the stars with extrasolar planets by calculating the percentage difference between the temperature of the hottest star and that of the Sun and between the temperature of the coolest star and that of the Sun. Comment on your results.

2. Quantitatively compare the range in masses of the stars of extrasolar planets selected here by calculating the percentage difference between the mass of the most massive star and that of the Sun and between the mass of the least massive star and that of the Sun. Comment on your results.

Step 2—Temperature and Eccentricity

We can do some fairly simple calculations to find out how the temperature of a planet changes from its closest approach to its star during its orbit, periastron, to its farthest distance from its star during its orbit, apastron. We start by assuming that the amount of starlight the planet receives, the flux, is constant as is the amount of light absorbed and reflected by the planet. We don't know the axial tilts of the extrasolar planets, so we ignore any seasonal differences. We can then ask ourselves, "By what fraction or percentage does each extrasolar planet's surface temperature change from its perihelion to aphelion? How does this compare to similar changes on Earth?"

We know from the Stefan-Boltzmann law that the temperature is proportional to the amount of flux to the 1/4 power. We also know that the flux from a star decreases according to the inverse square law. We will work with ratios to find out the fractional or percentage change for periastron versus apastron. **Figure 30.1** shows comparisons of the temperatures at periastron versus the temperatures at apastron.

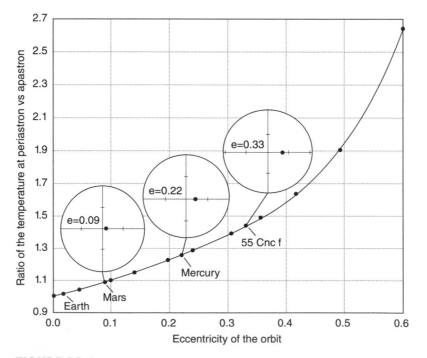

FIGURE 30.1

The Earth's eccentricity of 0.0167 is very close to 0 but not exactly at 0. The ratio of the perihelion temperature to aphelion temperature is about 1.017, or about 1.7 percent. The date of perihelion ranges between January 2 and 5; aphelion, between July 3 and 6.

As another example, with an eccentricity of 0.09, the temperature of Mars would be about 10 percent greater at perihelion than at aphelion. Assume that the axial tilt of the Earth does not change, nor does its average distance from the Sun.

3. What would the seasons be like if Earth's orbit had the eccentricity value of Mars?

4. Would the Northern Hemisphere's summer be longer or shorter? Explain.

5. Would the Northern Hemisphere's summer be hotter or colder? Explain.

6. From the graph of Figure 30.1, state the ratio of temperatures for Mercury. What percentage does this correspond to?

7. What would life on Earth need to do to withstand this temperature range if Earth's orbit had the same eccentricity as that of Mercury? Provide a few examples.

Step 3—Habitable Zones

A common definition of the habitable zone is that it is the range of distances from the central star in which liquid water might exist on the surface of a planet if the planet has a dense enough atmosphere. As under these criteria Earth is habitable, the Sun must have a habitable zone as well. One estimate of the Sun's habitable zone is shown in **Figure 30.2**. The light-gray shaded areas depict the conservative estimate of the habitable zone, while the dark-gray shaded areas show more optimistic estimates of the inner and outer boundaries of the habitable zone. The optimistic estimate includes an area about halfway between the orbits of Earth and Venus (inner edge) and the distance of Mars's orbit at the planet's aphelion. The conservative estimate starts just inside Earth's orbit and extends out to Mars's orbit at the planet's perihelion distance.

8. The full details of the orbits of the selected extrasolar planets are listed in **Table 30.2**, along with the conservative and optimistic estimates of the habitable zones. The sketches of the estimated habitable zones for the six extrasolar planets are shown in **Figure 30.3**. Four of the extrasolar planets have orbits that lie completely within the conservative estimate for the habitable zone of their star. Which planets are these?

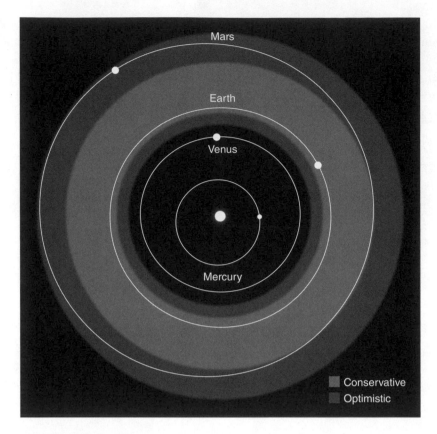

FIGURE 30.2

⊙ TABLE 30.2

Habitable zones of selected extrasolar planets. Identifying letters refer to the identification of the planetary system in Figure 30.3.

	PLANET DESIGNATION	DISTANCE FROM STAR (AU)	PERIOD (DAYS)	ORBITAL ECCENTRICITY	CONSERVATIVE (AU)		OPTIMISTIC (AU)	
					INNER HABITABLE ZONE	OUTER HABITABLE ZONE	INNER HABITABLE ZONE	OUTER HABITABLE ZONE
	Earth	1.00	365	0.06	0.95	1.4	0.85	1.7
(a)	HD 10180 g	1.42	602	0.00	1.13	1.94	0.87	2.11
(b)	HD 99109 b	1.11	439	0.09	0.82	1.43	0.62	1.56
(c)	HD 28185 b	1.02	379	0.05	0.95	1.65	0.73	1.80
(d)	HD 73534 b	3.02	1770	0.07	2.14	3.78	1.61	4.13
(e)	HD 183263 b	1.49	626	0.36	1.15	1.97	0.88	2.15
(f)	55 Cnc f	0.77	261	0.32	0.77	1.35	0.59	1.48

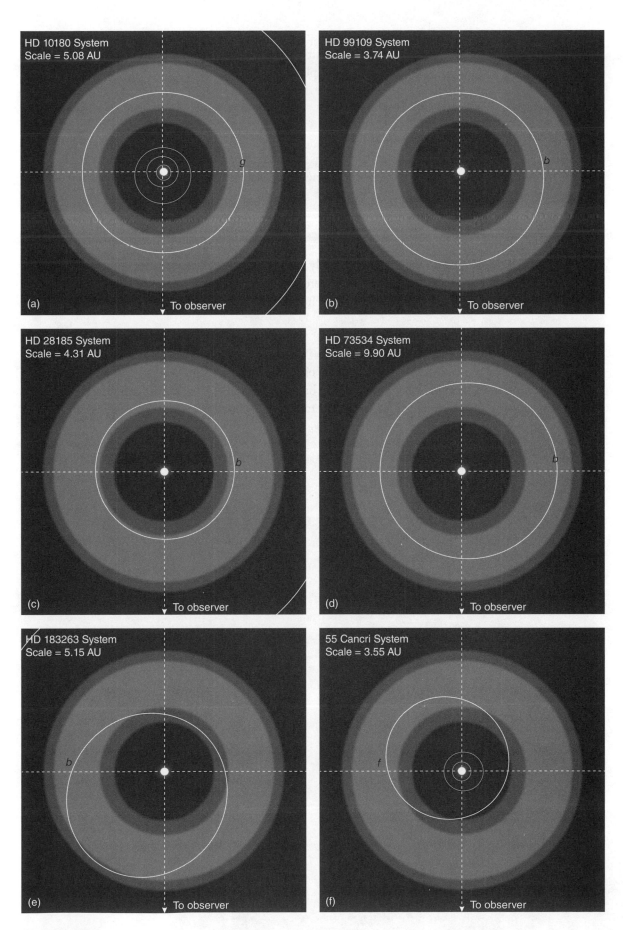

FIGURE 30.3

9. The extrasolar planets listed in Tables 30.1 and 30.2 of this activity were selected because of their low orbital eccentricities and their positions within the habitable zones of their respective stars. Which one of the four extrasolar planets that you listed above is most like Earth? In what ways is it different?

10. Earth's seasons and weather are not strongly affected by our orbital eccentricity. Which one of these extrasolar planets probably experiences the largest range in temperatures based on its orbit? How would it be affected and why?

11. Most of these extrasolar planets are probably gaseous, especially those close to Jupiter's mass or larger. If all of these extrasolar planets were to have at least one moon with a solid surface, which extrasolar planet's moon would you choose to visit or live on? Explain your decision.

12. Summarize what it would be like to live on either an extrasolar planet or a moon of an extrasolar planet that you concluded would be most habitable. Use the information about its distance from its star, its orbit, its mass and radius (if estimated), and its probable composition, stating in your description whether it is a gaseous or terrestrial planet.

Credits

Activity 1
Figure 1.5 (a–c): Courtesy of Ana M. Larson, University of Washington, Astronomy Department.

Activity 4
Figure 4.3 (a): Courtesy of Ana M. Larson, University of Washington, Astronomy Department; Fig. 4.4 (a–k): Shutterstock.

Activity 9
Figure 9.1 (from left to right): NASA E/PO, Sonoma State University, Aurore Simonnet; NASA/CXC/SAO; NASA/JPL-Caltech; © Laurie Hatch; NASA/JPL/Caltech Joint Astronomy Center in Hilo, Hawaii; NRAO VLA Image Gallery; The National Radio Astronomy Observatory, Green Bank; Dave Finley, Courtesy National Radio Astronomy Observatory and Associated Universities, Inc.

Activity 12
Figures 12.1, 12.3, 12.4: Courtesy of Ana M. Larson, University of Washington, Astronomy Department.

Activity 13
Figure 13.1: NASA/JPL; Fig. 13.3: NOAA.

Activity 14
Figures 14.1-14.2: NASA/JPL/Space Science Institute.

Activity 15
Figure 15.1 (top): NASA; Fig. 15.1 (bottom): NOAA; Fig. 15.2 (all): NASA/JPL/Space Science Institute; Fig. 15.3 (left): USGS and NASA; Fig. 15.3 (right): NASA/JPL/Caltech/ASI; Fig. 15.4: NASA/JPL/USGS; Fig. 15.5: NASA/JPL/Universities Space Research Association/Lunar & Planetary Institute.

Activity 16
Table 16.1–Table 16.2 (all): Courtesy of Ana M. Larson, University of Washington, Astronomy Department; Fig. 16.1 (b): Courtesy of Ana M. Larson, University of Washington, Astronomy Department.

Activity 19
Figure 19.1 (all): SOHO (ESA & NASA).

Activity 20
Figure 20.2 (both): NASA, NOAO, ESA and The Hubble Heritage Team (STScI/AURA).

Activity 21
Figure 21.1(a): NASA/CXC/SAO/F. Seward et al; (b): Palomar Observatory; (c): NASA/JPL-Caltech/R. Gohrz; Spitzer Space Telescope; (d): NRAO/AUI/NSF; Fig. 21.2: Jeff Hester and Paul Scowen (ASU), and NASA/ESA; Fig. 21.3: Jeff Hester and Paul Scowen (ASU), and NASA/ESA.

Activity 26
Table 26.1 and Fig. 26.1 (all): Palomar Observatory Sky Survey/NASA/AURA.

Activity 28
Figure 28.1: R. Williams (STScI), the Hubble Deep Field Team and NASA.

Index